EMOTIONAL GEOGRAPHIES

Emotional Geographies

Edited by

JOYCE DAVIDSON
Queen's University, Ontario, Canada

LIZ BONDI
Edinburgh University, UK

MICK SMITH
Queen's University, Ontario, Canada

ASHGATE

Published by
Ashgate Publishing Limited
Gower House
Croft Road
Aldershot
Hampshire GU11 3HR
England

Ashgate Publishing Company
Suite 420
101 Cherry Street
Burlington, VT 05401-4405
USA

Ashgate website: http://www.ashgate.com

British Library Cataloguing in Publication Data
Emotional geographies
  1.Spatial behavior 2.Emotional conditioning 3.Emotions
  4.Human geography
  I.Davidson, Joyce, 1971- II.Bondi, L. (Liz) III.Smith,
  Mick, 1961-
  304.2

Library of Congress Control Number: 2005924870

Paperback edition published 2007 / Hardback edition reprinted 2007

ISBN 978 0 7546 4375 3 (Pbk)   ISBN 978 0 7546 7107 0 (Hbk)

Printed and bound in Great Britain by MPG Books Ltd. Bodmin, Cornwall.

Contents

List of Tables and Figures

List of Contributors

Amanda Bingley is a Research Associate in the Institute for Health Research at Lancaster University. Following her doctoral research on the influence of gender on landscape perception she has continued to focus on the relationship between mental health and well-being and place. Amanda is currently involved in Palliative Care research with the International Observatory on End of Life Care. She has published in *Social and Cultural Geography, Social Science and Medicine* and is co-author of the book *Subjectivities, Knowledges and Feminist Geographies* (Rowman and Littlefield, 2002).

Liz Bondi is Professor of Social Geography at the University of Edinburgh. Informed by her longstanding involvement in feminist geography, her current research focuses on counselling and psychotherapy as socio-spatial practices, and on emotional geographies. Founding editor of the journal *Gender, Place and Culture*, she has published chapters in several edited collections, and numerous journals such as *Antipode, Environment and Planning A, Progress in Human Geography, Social and Cultural Geography,* and *Society and Space*. She is co-author of *Subjectivities, Knowledges and Feminist Geographies* (Rowman and Littlefield, 2002), and co-editor (with Joyce Davidson) of a special issue of *Gender, Place and Culture* (2004, 11: 3) on emotional geographies and co-editor (with Nina Laurie, of a special issue of *Antipode* (2005, forthcoming) on professionalisation, political action and neo-liberalism.

Nicola Burns is a Research Fellow at the Centre for Sentencing Research, Law School, University of Strathclyde. Since completing her doctoral research on the experiences of disabled people in the housing system, Nicola has worked on a variety of research projects including an Economic and Social Research Council project looking into the experiences of people with mental health problems in remote rural communities with Hester Parr and Chris Philo. Her research interests include housing inequality, disability, qualitative research methods and the criminal justice system.

Marion Collis is a Lecturer in Sociology at Monash University, Australia. She works at the university's Gippsland Campus in regional Victoria where she heads a small research unit which undertakes applied social and community research with a regional/local focus. Informed by a feminist sociology, her research interests include women's mental and reproductive health, and the impact of shiftwork on marital relationships and family life. She teaches the Sociology of Health and Illness, Women's Sociology and the Sociology of Reproduction.

David Conradson is Lecturer in Human Geography at the University of Southampton. Following his doctoral work on voluntary welfare organisations, which included examination of the experiential texture of various service spaces, he has become interested in notions of therapeutic environment and spacings of subjectivity more generally. His work on these matters has been published in *Environment and Planning A, Social and Cultural Geography* and *Health and Place*. He has edited a special issue of *Social and Cultural Geography* on Geographies of Care (2003, 4: 4) and is co-editor, with Christine Milligan, of a forthcoming volume entitled *Landscapes of Voluntarism: New Spaces of Health, Welfare and Governance* (Policy Press, 2006).

Joyce Davidson is Assistant Professor of Geography at Queen's University, Kingston, Ontario. Following her UK based doctoral research on agoraphobia, which formed the basis of *Phobic Geographies* (Ashgate, 2003), she has developed a research and teaching programme focused around geographies of health, embodiment and emotion. Organiser of the first interdisciplinary conference on Emotional Geographies (Lancaster University, September 2002) she has co-edited special issues on this subject for *Gender, Place and Culture* (with Liz Bondi, 2004, 11: 3) and *Social and Cultural Geography* (with Christine Milligan, 2004, 5: 4). She has published in sociology and philosophy as well as geography journals, and is co-author of *Subjectivities, Knowledges and Feminist Geographies* (Rowman and Littlefield, 2002).

Anthony Gatrell is Dean of the Faculty of Arts and Social Sciences and Professor of the Geography of Health at Lancaster University. For eight years he was Director of the Institute for Health Research, which was created in 1996. He also directed, for four years (2001-04), Health R&D NoW, the NHS R&D Support Unit in North-West England (a collaboration between Lancaster, Liverpool and Salford Universities). His research interests lie primarily in geographical epidemiology and the geography of health care provision, but with an underlying interest in health inequalities. He published *Geographies of Health: an Introduction*, Blackwell, 2002 and is the author or editor of three other books: *GIS and Health*, Taylor and Francis, 1998 – edited with Markku Loytonen; *Interactive Spatial Data Analysis*, Addison Wesley Longman, 1995 – co-authored with Trevor Bailey; and *Distance and Space*, Oxford University Press, 1983.

Jennie Germann Molz is an Economic and Social Research Council postdoctoral fellow in the Centre for Mobilities Research at Lancaster University in Lancaster, England. Following her PhD research on round-the-world travel websites, her current research focuses on the intersections between mobility, technology and sociality with particular interest in travel and tourism. She has published in several edited collections including *Tourism Mobilities* (Routledge, 2004) and *Culinary Tourism* (University of Kentucky Press, 2004) and in journals such as *Citizenship Studies* and *Environment and Planning A.*

Colleen Heenan is Senior Lecturer in Psychology at Bolton Institute, UK and adult psychotherapist in private practice. She was a co-founder of the Leeds (UK) Women's Counselling and Therapy Service. Her area of interest and research is gender, psychoanalysis and post-modern thinking, with particular reference to women, bodies and eating problems and she has edited a number of special features in the journal Feminism and Psychology as well as contributing to other texts and journals on these subjects. Colleen is co-author of two books (with Erica Burman *et al.*) – *Challenging Women: Psychology's Exclusions, Feminist Possibilities* (Open University Press, 1996) and *Psychology, Discourse, Practice: From Regulation to Resistance* (Taylor and Francis, 1996). She is also co-editor (with Bruna Seu) of *Feminism and Psychotherapy: Reflections on Contemporary Theories and Practices* (Sage, 1998).

Mike Hepworth is Honorary Reader in Sociology at the University of Aberdeen, By Fellow of Churchill College, Cambridge, and holds honorary appointments at the Universities of Sheffield and Abertay Dundee. His main research interest is the role of visual and literary images in the social construction of ageing/old age and he has published widely in this field. His most recent book is *Stories of Ageing* (Open University Press, 2000). He is a founder member of the editorial boards of *Theory, Culture and Society* and *Body and Society*.

Jenny Hockey is Professor of Sociology at the University of Sheffield. Her research interests span two fields: death, dying and bereavement; and gender studies. She has developed both these areas of scholarly activity through a variety of forms of collaborative work and she has held several Economic and Social Research Council grants. She is co-author of several books, including *Social Identities across the Life Course* (Palgrave, 2003); *Death, Memory and Material Culture* (2001 Berg); *Beyond the Body: Death and Social Identity* (Routledge, 1999); *Exploring Self and Society* (Macmillan, 1998), and co-editor of several others, including *Grief, Mourning and Death Ritual* (Open University Press, 2001); *Ideal Homes: Social Change and Domestic Life* (Routledge, 1999) and *Death, Gender and Ethnicity* (Routledge, 1997).

Phil Hubbard is Reader in Urban Social Geography and has a particular interest in the everyday geographies of the city. He has written extensively on sex work in the neoliberal city as well as the reconfiguration of urban governance. His books include *Sex and the City* (Ashgate, 1999); *People and Place* (Pearson, 2001); *Key Thinkers on Space and Place* (Sage, 2004) and the forthcoming *Key Concepts in Geography – The City*.

Owain Jones is an Associate Lecturer and course-writing consultant for the Open University, and a visiting lecturer at the University of Bristol and the University of the West of England. Since completing his doctoral research in childhood and rurality, he has conducted research in the School of Geographical Science, University of Bristol, in the areas of geographies of childhood and geographies of nature, place and landscape. He has published numerous papers in a range of journals, and a book, *Tree Cultures*, with Paul Cloke (Berg, 2001). He is Associate Editor of the *Children's Geographies* Journal.

Christine Milligan is a Lecturer in the Institute for Health Research at Lancaster University. Her research interests include informal caring and older people; mental health; voluntarism and social welfare; and therapeutic landscapes. Christine has written a book on *Geographies of Care: Space, Place and the Voluntary Sector* (Ashgate, 2001) and co-edited two further volumes: *Landscapes of Voluntarism: New Spaces of Health, Welfare and Governance* (Policy Press, forthcoming June 2006) and *Celtic Geographies* (Routledge, 2001). She has co-edited a special issue on emotional geographies for *Social and Cultural Geography* (with Joyce Davidson, 2004, 5: 4) and has published widely in international refereed journals such as *Social Science and Medicine; Health and Place; the Journal of Social Policy; Progress in Human Geography, Area* and *Environment and Planning A and C.*

Sara M. Morris is based in the Institute for Health Research at Lancaster University, UK. After a decade of conducting sociological research with people with cancer and their informal and formal carers, Sara is now working with service user and carer groups to support participative lay involvement in health research. Sara's cancer-focused work has been published in a wide variety of academic, practitioner and methodological journals, such as *Health, Risk and Society, Qualitative Health Research, European Journal of Cancer Care* and *Social Science and Medicine.*

Hester Parr is Reader in Human Geography at Dundee University and has interests in geographies of mental health, illness and disabilities. She is co-editor of *Mind and Body Spaces: New Geographies of Illness, Impairment and Disability* (Routledge, 1999) and author of the *Progress Reports on Medical Geography 2000-2003* (Progress in Human Geography). She has published widely on questions of mental health and social space in *Environment and Planning D: Society and Space, Transactions of the Institute of British Geographers, Social and Cultural Geography* and *Area*. She is co-editor with Chris Philo of theme issues on 'psychoanalytic geographies' (*Society and Cultural Geography*, 2003) and institutional geographies (*Geoforum*, 2000). She currently holds an Economic and Social Research Council Fellowship (2003-2007) concerned with 'Embodied geographies of inclusion: placing difference'.

Mark Paterson is Lecturer in Philosophical Studies and Cultural Studies at the University of the West of England, Bristol. His doctoral research investigated the tactile content of spatial experience, and since then he has been concerned more generally with the relationship between the sensory and the affective in embodied experience. He has published in journals such as *Angelaki: Journal of the Theoretical Humanities*, and written chapters in books such as *The Book of Touch* (Ed. Constance Classen, Oxford: Berg) and *Smell Culture* (Ed. Jim Drobnick, Oxford: Berg). Currently he is writing a book for Routledge, *Consumption and Everyday Life* (forthcoming, 2005).

Bridget Penhale is Senior Lecturer in the Department of Community, Ageing, Rehabilitation, Education and Research, within the School of Nursing and Midwifery at the University of Sheffield. She has a background in social work and her research interests include elder abuse and adult protection, bereavement and the mental health of older people. She is co-author of *Institutional Abuse: Perspectives Across the Life Course* (Routledge, 1999) and *Reviewing Case Management for Older People* (Jessica Kingsley, 1996). She has also published in a wide range of academic journals including *Ageing and Society, Journal of Interprofessional Care* and *Journal of Adult Protection*.

Chris Philo is Professor of Human Geography at the University of Glasgow. His major research interests have been on the historical geography of madness and asylums, and he is author of *A Geographical History of Institutional Provision for the Insane from Medieval Times to the 1860s in England and Wales: The Space Reserved for Insanity* (Edwin Mellen Press, 2004). He is also co-author of several major human geography texts including *Approaching Human Geography* (Paul Chapman, 1991) and *Practising Human Geography* (Sage, 2004); and co-editor of *Body Cultures: Essays on Sport, Space and Identity* (Routledge, 1998); *Entanglements of Power* (Routledge, 1999); and *Animal Spaces and Beastly Places* (Routledge, 2000). He is co-editor with Hester Parr of theme issues on 'psychoanalytic geographies' (*Society and Cultural Geography*, 2003) and institutional geographies (*Geoforum*, 2000); with Joe Painter of theme issues on 'spaces of citizenship' (*Political Geography Quarterly*, 1995); with Deborah Metzel of theme issues on 'geographies of intellectual disability' (*Health and Place*, 2005); with Fiona Smith of theme issues on 'political geographies of children and youth' (*Space and Polity*, 2003); and with Jennifer Wolch on 'post-asylum geographies' (*Health and Place*, 2000).

Until his retirement in 2003, **David Sibley** was Professor of Geography at the University of Hull. Immediately after retiring he spent a year as a Visiting Fellow at the National Institute for Spatial and Regional Analysis at the National University of Ireland, Maynooth. His research interests include the application of psychoanalysis to social problems, the social production of knowledge and the state's response to nomadic and semi-nomadic groups. He is joint editor of the journal *Social and Cultural Geography*. He has published widely and his books include *Geographies of Exclusion* (Routledge, 1995).

Mick Smith is Associate Professor of Philosophy and Environmental Studies at Queen's University, Kingston, Ontario. His current research is focused on questions of environmental responsibility and he has published widely in journals such as *Environmental Ethics, Environmental Values, Ethics, Place and Environment* and *Environmental Politics*. He is author of *An Ethics of Place* (SUNY, 2001) and co-author of the *Ethics of Tourism Development* (with Rosaleen Duffy, Routledge, 2003).

Deborah Thien is a feminist scholar with a long-standing interest in how gender, place, and culture intersect, particularly in terms of women's emotional health and well-being. She has recently completed a PhD in Human Geography at the University of Edinburgh. Her doctoral work examines geographies of emotion, combining geographical understandings of social and cultural places and spaces, feminist and critical theories of gender, and psychoanalytically-inspired insights into issues of self, identity, and relationality. The substantive focus of this research is the emotional well-being of women in Shetland, Scotland. Deborah has co-edited and contributed to an international collection of essays, *Geography and Gender Reconsidered* (Women and Geography Study Group, 2004) and has contributed a chapter to *Critical Studies in Rural Gender Issues* (Ashgate, 2004).

Carol Thomas is a Senior Lecturer in the Sociology of Health and Illness, based in the Institute for Health Research at Lancaster University. She has directed two large NHS-funded research projects in cancer and palliative care, one on the place of death preferences of terminally ill cancer patients and their carers, the other on the psychosocial needs of cancer patients and carers. Many publications have resulted in cancer and social science journals, including *Social Science and Medicine*. Her other research interests lie in Disability Studies. She is the author of *Female Forms: Experiencing and Understanding Disability* (Open University Press, 1999), and co-editor of *Disabling Barriers – Enabling Environments*, 2nd edition (Sage, 2004), and has published many journal papers and book chapters on disability themes. She is a member of the Executive Editorial Board of the journal *Disability and Society*.

John Urry is Professor of Sociology and Director of the Centre for Mobilities Research at Lancaster University. He has published various books including *Economies of Signs and Space* (1994, with S. Lash), *Consuming Places* (1995), *Contested Natures* (1998, with P. Macnaghten), *Sociology Beyond Societies* (2000), *Bodies of Nature* (2001, with P. Macnaghten), *The Tourist Gaze* (1990/2002) and *Global Complexity* (2003). He is about to establish a new journal *Mobilities* (with K. Hannam and M. Sheller). John Urry chaired the UK's Research Assessment Panel in Sociology in 1996 and 2001.

Chapter 1

Introduction:
Geography's 'Emotional Turn'

Liz Bondi, Joyce Davidson and Mick Smith

Clearly, our emotions *matter*. They affect the way we sense the substance of our past, present and future; all can seem bright, dull or darkened by our emotional outlook. Whether we crave emotional equilibrium, or adrenaline thrills, the emotional geographies of our lives are dynamic, transformed by our procession through childhood, adolescence, middle and old age, and by more immediately destabilising events such as birth or bereavement, or the start or end of a relationship. Whether joyful, heartbreaking or numbing, emotion has the power to transform the shape of our lives, expanding or contracting our horizons, creating new fissures or fixtures we never expected to find. But how do we articulate and negotiate such complex emotional landscapes?

On the surface, the discipline of geography often presents us with an emotionally barren terrain, a world devoid of passion, spaces ordered solely by rational principles and demarcated according to political, economic or technical logics (Parr forthcoming). But this apparent absence is hardly surprising since emotions are never simply surface phenomena, they are never easy to define or demarcate, and they not easily observed or mapped although they inform every aspect of our lives. Perhaps it would be better to say that geography, like many of its disciplinary siblings, has often had trouble expressing feelings. The difficulties in communicating the affective elements at play beneath the topographies of everyday life have meant that, to a greater or lesser extent, geography has tended to deny, avoid, suppress or downplay its emotional entanglements.

This is beginning to change, as the recent appearance of publications, conference sessions and courses dedicated to the subject of emotion demonstrates. A new interest in, and upsurge of, emotion is evident in writings about people and places, and surely signifies more than a passing academic fad. This emerging body of work is obviously critical of past presuppositions that emotions are not materially important. However, it also tries to recover something of those aspects of geographical and allied traditions that have implicitly, if not always explicitly, acknowledged the presence of emotions in our interpretations and understandings of the world. Thus, perhaps the recent 'emotional turn' in geography results as much from positive recognition that emotions *already have* an important place in our own and others' work, as from any sudden appearance of a shiny new 'object' of study.

There is also a feeling, embodied in, for example, Kay Anderson and Susan Smith's (2001) influential guest editorial in the journal *Transactions*, that emotions have an important role to play in maintaining geography's critical edge. An academic world that is increasingly business and policy driven suffers constant pressures to quantify and make economically tangible its subject matter. Although emotions can certainly be manipulated, managed and perhaps even manufactured for commercial and political purposes (Hochschild, 1983; Mestrovič, 1997), their subterranean ebbs and flows also resist attempts to represent them simply as untapped social (or academic) resources. A genuine emotional geography cannot just deal in feelings, like a stockbroker deals in dollars, or measure policy outcomes in terms of some bureaucratically derived hedonistic calculus. It must try to express something that is ineffable in such objectifying languages, namely a sense of emotional involvement with people and places, rather than emotional detachment from them. All of the chapters in this book attempt to do this, although the methods they employ and their modes of expression differ considerably.

Our own positions as editors of this collection merit brief comment. Such accounts are never entirely straightforward, but we can point out that for various reasons, both personal and political, and in various ways, both theoretical and empirical, we have all in recent years become increasingly involved in the practice of emotional geographies. For each of us, our work is infused with and informed by emotion in complex and dynamic ways. We are, like the majority of academic researchers, emotionally committed to our work. That is to say, we care deeply about the subject(s) of our research. Additionally, for each of us, the focus of our work is, in some senses at least, *emotional* in nature. Whether specifically, as in spatially mediated *angst* and *fear* of places deemed threatening because they are 'peopled' (Davidson 2003), or environmentally 'induced' *awe* and *love* for a 'natural' world felt inherently worthy of protection and respect (Smith 2001), or for the seemingly inexhaustible ensemble of disruptive emotional experience, from concern about the success or stress of a career, to unbearable loss of and longing for a loved one, that leads many to psychotherapies whether as practitioners or recipients (Bondi 2003a; Bondi with Fewell 2003). Our various research projects are thus motivated and informed by an understanding that emotions are situated within, and co-constitutive of, our working (as well as social) lives. However, like many others it is only in recent years that we began to conceptualise our work and the connections between our interests in terms of emotional geographies.

Our coming together as editors, and our joining with contributors, to produce this collection, expresses a more widely felt interest and need. There is, as this book shows, a growing desire for and commitment to the project of placing emotions less peripherally in our research and writings, in terms of both individual projects and broader (inter)disciplinary concerns. We hope to demonstrate that a spatially engaged approach to the study of emotions is capable of bringing new insights to geographical research. Moreover, despite the superficially reductive title of the collection, we aim to undermine any rigid adherence to and defence of disciplinary boundaries that characterises much contemporary academic enterprise. The term 'emotional geographies' should not be understood narrowly since emotions slip through and between disciplinary borders. This is not a new subdiscipline of an already established field, since we have previously remarked that geography has largely defined itself in terms that exclude the emotional. Rather, as many of the chapters gathered here illustrate, despite their disciplinary differences, there remains a common concern with the spatiality and temporality of emotions, with the way they coalesce around and within certain places. Indeed, much of the symbolic importance of these places stems from their emotional associations, the feelings they inspire of awe, dread, worry, loss or love. An emotional geography, then, attempts to understand emotion – experientially and conceptually – in terms of its socio-*spatial* mediation and articulation rather than as entirely interiorised subjective mental states.

In order to illuminate this argument, we introduce three core themes through which this volume is organised, namely the location of emotion in both bodies and places, the emotional relationality of people and environments, and representations of emotional geographies. In relation to each we identify antecedents to the emergence of emotional geographies and then introduce the chapters in the corresponding section of this volume. Many chapters engage with more than one of these themes, and our introductory overview seeks to highlight some of these cross-linkages.

The first source of inspiration we discuss stems from critical geographies of health and embodiment. Without focusing specifically on emotion, such work has illuminated *where* emotions are felt to reside, notably in both bodies and places. Picking up on this theme the chapters that form the first main section of this volume engage explicitly with questions of where and how emotions are located through studies that extend geographies of health and embodiment in new directions. The second source of inspiration we discuss includes geographies of identities and social relations. Again, although such work does not focus directly on emotion, it highlights how emotions are produced in relations between and among people and environments. The relationality of emotion, illustrated in the first section of this book is explored in greater depth in the second section through studies that consider the relationality of emotions across a range of spatial scales and contexts. The third source of inspiration on which we draw are theoretical perspectives that problematise and facilitate the representation of emotion. In the third section of the book we include chapters that analyse representations of emotion and that experiment with modes of emotional representation.

Our account of key antecedents of emotional geographies is necessarily partial and tentative, and we offer it to draw out what we consider to be instructive or illustrative themes and examples that are emotionally poignant and powerful, effective as well as affective, and in relation to which emotional geographies can usefully develop. In the course of this account we argue for a non-objectifying view of emotions as relational flows, fluxes or currents, in-between people and places rather than 'things' or 'objects' to be studied or measured. In so doing we hope to give the reader a feel(ing) for the spatiality of emotion elaborated in the chapters that follow.

Locating Emotion

One of the specialisms of geography that has been most willing to admit emotion into its production of knowledge is that of critical geographies of chronic illness and disability. In this field, researchers have explicitly recognised the importance of understanding and faithfully representing the emotional experiences of those they study. For example, in her study of the workplace experiences of women with multiple sclerosis (MS), Isabel Dyck (1999) highlighted individuals' feelings of exclusion and oppression, and their sense of struggling with their impairments and symptoms. The emotionality of such experience is particularly evident in attempts to renegotiate social relations in the context of the stigmatising – disabling – attitudes often found in workplace environments. Pamela Moss (1999) has also examined the emotional turmoil of negotiating ill-health in places of work, drawing on her own experience of myalgic encephalomyelitis (ME), and reporting a complex range of responses including compassion and understanding from some, but dismissive, antagonistic and coercive behaviours from others, who positioned her as overly emotional, unable to handle stress, and as suffering from 'psychosomatic' and therefore somehow 'unreal' physical symptoms. In a similar vein, which she explicitly and unapologetically refers to as 'depressing', Vera Chouinard (1999a, 270) has documented her own and others' experiences of disability.

Research concerned with mental ill-health has extended these pioneering engagements with emotion in additional directions. Several studies have illuminated the emotional experiences associated with symptoms of mental health conditions including agoraphobia (Bankey 2002; Davidson 2003), specific phobias (Davidson and Smith 2003), obsessive compulsive disorder (Segrott and Doel 2004) and psychotic illnesses (Parr 1999), while others have explored the complex emotional impacts of the deinstitutionalisation of medical care for the mentally ill (Milligan 1999; Parr 2000; Kearns and Gleeson 2001). In this body of literature, illnesses and symptoms interweave with caring and careless environments to produce complex and sometimes confusing emotional geographies negotiated by sufferers, carers and others. Karen Dias (2003), for example, has presented a powerful and sometimes shocking account of anorexic experience as it is narrated by its 'owners' in cyber-space. Cyber-space, Dias suggests, can offer a safer, less confrontational place for the expression of emotional pain and continual struggle so characteristic of eating disorders. Her respondents experience fleeting yet powerful feelings of accomplishment through (self-)denial, but are also locked into a place of loneliness, vulnerability, and desperation, rocked by desires to be 'normal' and understood rather than attacked by others. Their embodiment of emotion is deeply personal yet patterned and shaped by a sense of sharing with those others whose bodies are similarly placed.

In different ways, these studies locate emotions in 'othered' bodies, which are differently experienced in different places. But they do not suggest that emotions belong uniquely within, or are generated uniquely by, those suffering impairments, ill-health, or the burden of diagnostic categories. Thus, notwithstanding their focus on people categorised as unwell or disabled, in a variety of ways they point out that much is shared with those who are not so categorised. In some cases this is suggested through autobiographical accounts that mobilise commonalities between authors and colleagues (Chouinard 1999b; Moss 1999); in other cases accounts of 'disordered' emotional geographies are rendered legible through affinities to 'normal' everyday spatial experiences (Davidson 2003). Among such affinities are the psycho-social and material boundaries through which we differentiate ourselves from others and from our environments, and which ordinarily 'contain' or embody our emotions.

As geographies of embodiment have elaborated, bodily boundaries are frequently perceived and negotiated in emotionally powerful, disruptive and conflictual ways (Longhust 2001). The feelings of pride and pleasure, and/or guilt and shame bound up with dietary, exercise and cosmetic regimes reveal that our bodies are intensely emotional(ised) areas and thus an important focus for, and locus of, work on how and why, what and where we feel (Davidson and Milligan 2004; Grimshaw 1999). Responses to bodies of others considered transgressive reveal the strong feelings often provoked by fleshy boundaries. For example, Lynda Johnston (1996) found that the bodies of women bodybuilders often elicit horror and repugnancy because they are not only 'feminine' but also 'built'. Bodybuilding is one of many methods used to give form to feelings (and feelings to form), and other embodied geographies show the lengths to which many of us go to control or change our bodies (and our selves) frequently through practices of consumption (Bell and Valentine 1997; Colls 2004; Crewe 2001; Garvin and Wilson 1999). These studies also emphasise how embodied emotions are intricately connected to specific sites and contexts (Domosh 2001; Mathee 2004).

Questions about how emotions are embodied and located merit further elaboration in the context of typical and less typical everyday lives. This task is taken forward by the first main section of this volume, which develops our appreciation of the interconnected location of emotions in people and places. The chapters making up this section examine ordinary and extraordinary emotional experiences in domestic and institutional settings, indoors and out of doors, in places called home and places away from home. The section begins with a chapter concerned with experiences that are simultaneously extraordinary and very ordinary indeed. Sara Morris and Carol Thomas present a deeply moving account of the significance of place for people approaching death and those with whom they share the end of their lives. Through interviews with individuals who are terminally ill and with the often intimate others who care for them, the authors illuminate how the degenerative processes of advanced cancer leave sufferers 'infirm', and their bodies unbounded, whether in the spaces of 'private' homes or 'public' institutions. It seems that there is no 'right' place for such a death, but conclusions must be reached, and the authors explore how shifting embodied, social and spatial relations inform what is always an emotionally painful process of 'decision'.

Chapter three also engages with issues typically framed in terms of loss, but in the form of hysterectomy rather than end of life. This medical intervention is performed frequently and sometimes almost routinely on women, and yet very little is known about women's post-operative experiences. Drawing on interviews with 20 such women, Marion Collis presents an original analysis that privileges personal rather than medical(ised) accounts of hysterectomy. She highlights the importance of the social context of women's lives and their perceived social role for understanding the nature of their emotional responses. The narratives on which she draws are emotionally complex and fluid rather than fixed, and the chapter makes clear that the womb is not universally felt to be crucial to the 'performance' of femininity. By no means all women are emotionally attached to their wombs, and Collis suggests that emotional responses to hysterectomy vary according to factors such as employment – seen as a source of identity outside of motherhood – and across the life course. For example, young women tend to experience the removal of their womb as a 'tragedy', and grieve both for the loss of the child(ren) they might have had, and, if they are childless, for their anticipated identities as mothers. Rather than 'mourning the loss' older women often report that they have 'no regrets'.

In chapter four Christine Milligan, Amanda Bingley and Anthony Gatrell continue the theme of how emotions are embodied through an exploration of the shifting nature of emotional attachment to place among older people. Older people are routinely spatially marginalised often with profoundly hurtful, isolating and restrictive consequences. In contrast to this, the authors explore older people's positive emotional experiences of the shared community spaces of social and gardening clubs. Drawing on innovative qualitative research in these settings, they demonstrate that social spaces, and especially shared outdoor gardening activities, can play an important and constructive role in facilitating emotional expression and feelings of self-worth and belonging, in older people's lives.

The remaining two chapters in this section locate emotions in the bodies of tourists and in the sites they encounter on their travels. In chapter five, Jennie Germann Molz draws on websites published by North American travellers on trips around the world to analyse the mixed emotions these tourists describe in their accounts of eating at the global franchises of McDonald's. These narratives reveal how outlets of McDonald's are experienced and represented both in terms of comfort, familiarity and homeliness, and in terms of guilt and betrayal. Germann Molz examines the ingredients of this ambivalence, and shows how the apparently conflicting emotions map onto understandings of McDonald's spaces as alternately global and local, home and away.

Rounding off section one, in chapter six, John Urry brings the focus directly to the ways in which emotions are located in the constitution of places. Focusing on places consumed as tourist destinations, he shows how specific sites are constructed in ways saturated with emotion, sometimes wild and frightening, sometimes aesthetically pleasing and relaxing, sometimes dependent on ideas about rootedness, and so on. These emotions become integral to how places are imagined and portrayed, with profound implications for the embodied experiences of both tourists and residents.

In focusing on emotions experienced by embodied individuals and attaching to particular places, the chapters in section one all point to the inter-relatedness of people and their environments. It is to these inter-relations that we now turn.

Relating Emotion

As we have already observed, emotions are widely understood to be contained by the psycho-social and material boundaries through which embodied persons are differentiated from one another and from their surrounding environments. Although taken-for-granted in the course of much everyday life, such boundaries are never impermeable or entirely secure. Geographies of agoraphobia, for example, have shown how bodily boundaries are radically disrupted during experiences such as panic attacks, and how, long after the intensity of a panic attack subsides, the sufferers' boundaries often continue to feel extremely vulnerable and fragile (Bankey 2002; Davidson 2001). Much more welcome breaches of our psycho-social boundaries are suggested by experiences of 'being moved' by others, and by music, art, literature or landscapes: such experiences show how what appears to be outside impacts profoundly on our emotional interiors, getting *through* (across our boundaries) to us. In different ways, both 'disordered' and more ordinary emotional experiences highlight the permeability and fluidity of bodily boundaries. They also illuminate how emotions help to construct, maintain as well as sometimes to disrupt the very distinction between bodily interiors and exteriors. The close connections between boundary-forming processes and emotions suggest that it may be productive to think of emotions as intrinsically relational.

That social identities and inequalities are relational is widely accepted by social scientists: we define ourselves, at least in part, in terms of what we are not, and these relational claims are bound up with relations of inequality and oppression. Although highlighting the emotionally troubling hardships and injustices caused by inequalities and oppressions, researchers have not generally considered how emotions might underpin them. One important exception is David Sibley's (1995) influential account of *Geographies of Exclusion*, which advances an interpretation of the socio-spatial dynamics of racism and other oppressions that accords emotion a central role. For Sibley, geographies of exclusion can be understood as manifestations of conscious and unconscious feelings that arise in the real and imagined movement between 'selves' and 'others', expression of which is played out in the exclusionary qualities of social life. Several subsequent analyses have taken forward this perspective, arguing that attending to feelings of fear, anxiety, anger, envy and hatred is essential if we are to understand the insidious power and tenacity of racism (Kobayashi and Peake 2000; McKittrick 2000).

The importance of emotion is also suggested by studies of sexuality concerned with the oppression and repression of sexual dissidents. For example, in a challenge to geographers' traditional 'squeamishness' about the 'private' sphere of sexual, and especially 'scary' sexual feelings and actions such as fetishism and sado-masochism, Phil Hubbard (2000) has attempted to tease out some complexities of the moral(ising) geographies that underpin reluctance to engage with issues of desire and disgust in other areas of the discipline. He suggests that moral panics arise in response to behaviours (themselves expressions of feelings) that are considered somehow abnormal or wrong. Thus, the heteronormativity of social space, amply demonstrated by geographical research about gay and lesbian experiences (Valentine 1993; Duncan 1996), is fostered and sustained through the viscerality of homophobic emotional responses to such simple and ordinary acts as two same-sex adults holding hands (Valentine 1996). While homophobic disgust, hatred and hostility sometimes erupt into anti-gay violence (Myslik 1996), it more routinely makes its presence felt through judgemental moral outrage (Valentine 1996). The relationality of emotions contributes to the enforcement of other dominant stereotypes and norms including those of gender (Mehta and Bondi 1999; Namaste 1996; Browne 2004) and disability (Chouinard and Grant 1996; Chouinard 1997).

Critical geographers, including those working to counter racism and homophobia, have sought to undermine claims about 'naturalness' or 'inevitability' of social categories and spatial divisions. Exploring the relationality of emotion offers a promising avenue through which to advance understandings of dynamic geographies of difference, exclusion and oppression. But it is important also to attend to – and denaturalise – emotional geographies of connection, pleasure, desire, love and attachment. For example, the relationality of emotions that underpins the dynamics of exclusion also produces and sustains the psycho-social bonds of kinship and friendship. Feelings of desire, love and attachment, which have long been implicit in studies of familial and intimate relationships (Jamieson 1998) and geographies of sexuality (Bell and Valentine 1995), are increasingly emerging as a focus for explicit exploration. For example, Deborah Thien (2004) has sketched out a feminist geography of love, while Jaqui Gabb (2004) has examined the complex boundaries and overlaps between mother-baby and adult-sexual love. While these studies focus primarily on emotional dimensions of interpersonal relationships, other research has considered dynamic relationships between people and landscapes that implicitly mobilise emotions. For example, studies of therapeutic landscapes argue that places can be directly health promoting in ways that imply that particular kinds of environments have the capacity to transform people's (emotional) lives (Gesler 2003).

The chapters that make up the second section of this volume explore emotional aspects of relations between people and environments at different scales and in diverse contexts. Thus, the environment considered in chapter seven consists of a geographical region, while chapters eight, nine and ten focus on particular sites, and chapters eleven and twelve bring attention to the sensory interface between bodies, food and touch. These chapters engage with questions of how environments, variously conceived, are encountered as sources of distress, pleasure and commemoration, sometimes intensifying exclusion and sometimes fostering well-being.

Chapter seven is concerned with emotional geographies of the Scottish Highlands. As Hester Parr, Nicola Burns and Chris Philo elaborate, thinking through the Scottish Highlands as an emotional terrain might involve the use of signifying words like 'hard', 'tough', 'reserve' or 'repression', but also 'caring', 'closeness' and even 'romance' to describe the sociality of these peopled landscapes. These emotional topographies become poignantly meaningful in relation to the daily realities of Highland residents with mental health problems on whom the authors focus. Drawing on interviews with psychiatric service users, they examine how social relations involving emotional distress, recovery and care are configured in Highland localities. In so doing they offer important insights into the intensely and often deeply hidden emotional experiences produced in the interplay between and among people and environments. The authors conclude by considering complex issues of representing emotions, which we take further in the final section of this chapter and the book as a whole.

Chapter seven highlights how the unique and distinctive physical environment of the Scottish Highlands might ameliorate or exacerbate troubling emotions associated with mental health problems. Chapter eight takes up the idea that 'natural' environments might facilitate well-being (also suggested in chapter four). Focusing on encounters with a landscape of woods, heathland and shoreline offered within the context of a respite care centre, David Conradson draws out connections between qualities attributed to external spaces and to interior experiences narrated by some of the centre's guests, through which he advances a relational approach to the self-landscape encounter. Informed by the idea that particular kinds of landscape have therapeutic qualities, this contribution elaborates emotional geographies integral to such healing possibilities.

Chapter nine turns to the very different landscapes and emotions associated with 'nights out'. Phil Hubbard explores how evening economies evoke complex emotions mobilised in diverse relationships between people and commercialised leisure environments. Comparing people's perceptions and experiences of going out for the evening in city-centre and out-of-town venues, he reveals emotionally ambivalent responses to spaces typically cast as 'emotionally charged'. He shows that people often feel strongly about what it means to go out at night, and associate positive and pleasurable, or negative and unwanted emotions with different places at different times. Different kinds of emotional management are required depending on whether a 'quiet' or 'big' night out is considered desirable. City centres tend to be linked with feelings of anxiety and/or excitement, and are thought to be lively and stimulating, as opposed to the polite and predictable, relaxed and comfortable space of out-of-town leisure parks. Hubbard's exploration of these differences illustrates the dynamic emotional interplay between (simultaneously imagined, material and social) environments and people who variously use and avoid them.

Chapter ten returns to the theme of bereavement discussed in chapter two, also emphasising its extraordinary ordinariness, especially among the lives of older people. In a moving and poignant account, Jenny Hockey, Bridget Penhale and David Sibley explore some of the spatialised dimensions of emotional experiences associated with the loss of a heterosexual partner is later life. They show how domestic spaces and artefacts become 'environments of memory' within which the bereaved person continues to engage in an active relationship with his or her former partner. The authors show that the social presence of the dead is realised through everyday objects and practices, and their use of a case study approach reveals the operation of memorialising strategies through which persons, environments and emotions become inextricably entwined. They also highlight how these interweavings bear the impress of constructions of gender.

Chapter eleven considers one of the troubling 'discontents' associated with consumer culture, namely experiences of eating disorders. Eating is one of the ways in which we routinely incorporate cultural products into our bodies, thereby highlighting the permeability of bodily boundaries. The preparation and consumption of food are cultural practices saturated with emotions, including those of pleasure and love, but also those of guilt and anxiety. Moreover, as ideas such as 'comfort eating' highlight, everyday uses of food often seek to address desires that have little to do with nutritional needs. Drawing on experiences articulated in the context of a therapy group for women with eating disorders, Colleen Heenan offers a feminist psychoanalytic interpretation of the gendered relationships between people and the environments which women (and men) negotiate as they search for emotional and other forms of nourishment. Her account highlights the complex, confusing and unstable relationship between exterior and interior emotional geographies, epitomised by the idea of 'looking in the fridge for feelings'.

Perhaps because of the permeability of psycho-social boundaries, social norms set precise and restrictive limits on the skin to skin touch between individuals, which come into view in chapter twelve. Beyond familial contexts, many forms of touch – other than the mutual shaking of hands – are deemed inappropriate and transgressive, soliciting emotional responses of anxiety, fear or anger. However, touch may also be craved and is offered professionally by therapists such as Reiki practitioners. In the final chapter of the second section of this volume, Mark Paterson draws on personal experience of receiving Reiki massage, together with interviews with practitioners, to examine the nature of positive relations between tactile experience and affective state. For its practitioners, the Reiki – 'universal life force' in Japanese – is passed between people through touch. Touch thus becomes an irruptive, interpersonal experience that traverses skin and flesh, that cuts across Occidentally constructed divisions of mind and body, spiritual and emotional, outside and inside skin. The chapter thus demonstrates that the senses provide a pathway into profound emotional conditions, and how touch in particular can be both affective and expressive.

Throughout the chapters in sections one and two, authors attend to complex issues of how to represent emotions, which as we have observed have intrinsically ineffable qualities that render every representation lacking in some way. Emotional geographies therefore press against the limits of, and contribute to theoretical debates about, representation. It is to questions of representing emotions that we now turn.

Representing Emotion

One task for emotional geographies is to examine how modes of representation mobilise, produce and seek to shape emotions. Symbols of national pride, for example, are designed to foster forms of belonging that may render critical questioning problematic. Nichola Wood (2002), for example, has shown how music can be particularly effective and affective in generating and sustaining non-conscious, visceral emotional attachments, sometimes evoking sentiments almost despite ourselves, such as feeling inexplicably aroused by national anthems. The pleasures of 'losing oneself' in music thus co-exist with the dangers of how our lost selves might be exploited outside of our awareness. Understanding the emotional dimensions of artistic, political, and commercial representations in relation to their various spatial and temporal contexts is therefore a project of considerable importance and urgency for critical social scientists.

While emotions are powerfully mobilised in some kinds of representation, they also often elude the gaze of academic researchers and therefore fail to secure an appropriate place within the representations produced by academics. This has prompted some commentators to consider how to honour what remains unrepresented in the experiences, dynamics and very liveliness of everyday geographies (Thrift and Dewsbury 2000). One response has been to engage with non-discursive practices immediately recognisable as profoundly emotional, including, for example, movement and dance (McCormack 2003). But a conundrum remains: how can we represent that which lies beyond the scope of representation?

Of course correspondence to some objective reality is not possible in any domain: knowledge production – and representation – is always creative and contestable rather than neutral and reliable. Consequently issues of how to represent emotion call for those involved in generating emotional geographies to consider the emotion work done via the writing and reading of their texts as well as in their fields of study. A range of theoretical resources are available to support such thinking, ranging from phenomenological efforts to transcend distinctions between objective and subjective (Davidson 2003), through feminist engagements with reflexivity (Hertz 1997; Moss *et al.* 1993; Rose 1997), performative and non-representational approaches (Probyn 2003; Thrift 2004; but see Nash 2000), and psychoanalytic approaches to working with unconscious thoughts and feelings (Bondi 1999; Callard 2003). Some of these resources underpin contributions to this volume, but what we seek to highlight in this section are the genres within which we write and the concepts we deploy.

To a greater or lesser extent, academic writing articulates the preoccupations and passions of its authors, but traditional academic genres seek to exclude its explicit expression. Increasingly, however, academics have sought to experiment with more emotionally expressive styles. This section includes five chapters, each of which engages with emotional representations in different ways and in different voices. In some respects, chapters thirteen and fourteen remain bound within the traditional academic style of dispassionate commentary, and yet both seek to communicate a depth of emotional understanding with their different research subjects in ways that help to imbue such writing with expanded meanings. The remaining chapters all draw on autobiographical vignettes, whether to disrupt academic conventions (in chapter fifteen) or to evoke emotional experiences that illustrate the authors' arguments (in chapters sixteen and seventeen).

However problematic efforts to represent emotion might be, art forms including music and painting often seek to represent as well as to evoke emotions, and these representations provide useful insight into cultural constructions of emotion. In this context, chapter thirteen, by Mike Hepworth, elaborates a sociological analysis of the emotions and applies it to the way in which Victorian paintings frame the emotions of old age. Echoing scepticism about the idea that emotions of ageing and old age have qualities that distinguish them from emotions experienced during other periods of the life course also expressed in chapter four, Mike Hepworth examines how such claims are expressed, mediated, and given cultural frame and form. Focusing on paintings by two British artists popular in the Victorian period he explores how emotions of ageing are framed within the space of the canvas, together with the complex interplay between the intentions of the artist, artistic conventions of the time, and responses of the viewing public in relation to these representations of emotion.

In chapter fourteen, Deborah Thien explores the relationship between intimacy and distance, drawing on accounts of emotional well-being offered by women living in island communities in the north of the UK. She illustrates the relationality of emotions shaped through subtle and complex enactments that produce both people and the contexts within which they live. In so doing she also seeks to undo dominant assumptions about intimacy and to put into circulation meanings capable of containing emotional ambivalence and non-Euclidean geographies. This re-presentation of intimacy offers an original feminist elaboration of emotional geographies.

Chapter fifteen deploys autobiography to simultaneously explore and trouble representations of emotion. Through memory – conscious and unconscious, psychic and somatic – we all carry traces of past geographies, in ways that are always emotionally coloured in hues ranging from pale to vivid. In this context, Owain Jones emphasises that we are not, and cannot be, reflexively aware of, or in control of, how emotions are mapped onto us at the moments of our experience, or of how they are retained and retrieved (or not) through differing forms of memory. Moreover, he shows how emotions experienced as if in the present moment are never free of the past but are instead always re-encountered, in ways that simultaneously evoke familiarity and freshness. These living emotional depths of being frame our conscious selves and our rationality, so if we are to take account of emotions then we have to heed, as best we can, the force and direction of these processes.

In chapter sixteen, inarticulate emotional attachments to place are also evocatively highlighted by Mick Smith, who describes how the resurfacing of subterranean river waters in the land of his birth move him 'more than [he] can say'. He suggests that such attachments transcend and undo conventional distinctions between people and natural environments, and he explores how we might rethink and re-present our place within the world. Drawing on philosophies of meaning offered by Hans Georg Gadamer and Ludwig Wittgenstein, Smith seeks to develop an approach to environmental ethics that takes emotion seriously. Exploring and mobilising metaphors of ceaseless movement and geological time, he calls for us to open ourselves to feelings and meanings more usually silenced in academic writing.

The final chapter of this volume also calls for closer consideration of researchers' emotions. Liz Bondi explores the place of emotions and emotion work in processes of conducting research, testifying to the intense but often unacknowledged emotionality of researchers' experiences. She argues that much academic work could be enriched and strengthened if emotions were taken more seriously and placed less peripherally in relation to the academic enterprise. This is not a call to routinely write feelings into accounts of research in a researcher-centred way. Rather, the author suggests that heightened, reflexive awareness of researchers' emotions might inform the organisation and representation of academic research in a variety of ways, depending upon the underlying epistemologies of knowledge production. Perhaps this volume will go some way towards enabling such developments.

It is difficult to imagine any area of the social sciences or humanities that could not be enriched by the incorporation of the emotions that are so intricately entwined with the fabric of our lives. We hope that this volume will contribute to emotionality becoming less tangential to research in geography and its disciplinary siblings. We also hope that deeply geographical dynamics of emotion will be illuminated affectively and effectively in the chapters that follow.

References

Anderson, Kay and Susan Smith (2001) Editorial: emotional geographies, *Transactions of the Institute of British Geographers* 26: 7-10.

Bankey, Ruth (2002) Embodying agoraphobia: rethinking geographies of women's fear, in Liz Bondi, Hannah Avis, Amanda Bingley, Joyce Davidson, Rosaleen Duffy, Victoria Ingrid Einagel, Anja-Maaike Green, Lynda Johnston, Sue Lilley, Carina Listerborn, Mona Marshy, Shonagh McEwan, Niamh O'Connor, Gillian Rose, Bella Vivat and Nichola Wood, *Subjectivities, Knowledges, and Feminist Geographies.* Lanham MD: Rowman and Littlefield, pp. 44-56.

Bell, David and Gill Valentine (1997) *Consuming Geographies: You Are Where You Eat.* London: Routledge.

Bell, David and Gill Valentine (eds) (1995) *Mapping Desire. Geographies of Sexuality.* London: Routledge.

Bondi, Liz (1999) Stages on journeys: some remarks about human geography and psychotherapeutic practice *The Professional Geographer* 51: 11-24.

Bondi, Liz (2003) A situated practice for (re)situating selves; trainee counsellors and the promise of counselling, *Environment and Planning A* 35: 853-870.

Bondi, Liz with Judith Fewell. (2003) 'Unlocking the cage door': the spatiality of counselling, *Social and Cultural Geography* 4: 527-547.

Browne, Kath (2004) Genderism an the bathroom problem: (re)materialising sexed sites, (re)creating sexed bodies *Gender, Place and Culture* 11: 331-346.

Callard, Felicity (2003) The taming of psychoanalysis in geography, *Social and Cultural Geography* 4: 295-312.

Chouinard, Vera (1997) Making space for disabling differences: challenging ableist geographies *Environment and Planning D: Society and Space* 15: 379-387.

Chouinard, Vera (1999a) Body politics: disabled women's activism in Canada and beyond, in Ruth Butler and Hester Parr (eds) *Mind and Body Spaces: Geographies of Illness, Impairment and Disability.* London and New York: Routledge, pp. 269-294.

Chouinard, Vera (1999b) Life at the margins: disabled women's explorations of ableist spaces, in Elizabeth K. Teather (ed.) *Embodied Geographies: Spaces, Bodies and Rites of Passage.* London and New York: Routledge, pp. 142-156.

Chouinard, Vera and Grant, Ali (1996) On being not even anywhere near 'the project': ways of putting ourselves in the picture, in Nancy Duncan (ed.) *BodySpace: Destabilizing Geographies of Gender and Sexuality.* London and New York: Routledge, pp. 170-193.

Colls, Rachel (2004) 'Looking alright, feeling alright': emotions, sizing and the geographies of women's experiences of clothing consumption, *Social and Cultural Geography*, 5:4, 583-596.

Crewe, Louise (2001) The besieged body: geographies of retailing and consumption *Progress in Human Geography* 25:4: 629-640.

Davidson, Joyce (2001) 'Fear and trembling in the mall: women, agoraphobia and body boundaries', in Isabel Dyck, Nancy Davis Lewis and Sara McLafferty (eds) *Geographies of Women's Health.* London and New York: Routledge, pp. 213-230.

Davidson, Joyce (2003) *Phobic Geographies: The Phenomenology and Spatiality of Identity.* Aldershot: Ashgate.

Davidson, Joyce and Christine Milligan (2004) Embodying emotion, sensing space: introducing emotional geographies, *Social and Cultural Geography*, 5:4, 523-532.

Davidson, Joyce and Mick Smith (2003) 'Bio-phobias/Techno-philias: Virtual Reality Exposure as Treatment for Phobias of "Nature"', *Sociology of Health and Illness*, 25:6, 644-661.

Dias, Karen (2003) The ana sanctuary: women's pro-anorexia narratives in cyberspace, *Journal of International Women's Studies*, 4: 31-45. http://www.bridgew.edu/SoAS/jiws/April03/index.htm

Domosh, Mona (2001) The 'women of New York': a fashionable moral geography, *Environment and Planning D: Society and Space* 19: 573-592.

Duncan, Nancy (1996) Renegotiating gender and sexuality in public and private spaces, in Nancy Duncan (ed.) *BodySpace: Destabilizing Geographies of Gender and Sexuality.* London and New York: Routledge, pp. 127-145.

Dyck, Isabel (1999) Body troubles: women, the workplace and negotiations of a disabled identity, in Ruth Butler and Hester Parr (eds) *Mind and Body Spaces: Geographies of Illness, Impairment and Disability.* London and New York: Routledge, pp. 119-137.

Gabb, Jacqui. (2004) 'I could eat my baby to bits': passion and sexuality in mother-child(ren) love, *Gender, Place and Culture* 11: 399-415.

Garvin, Theresa and Kathleen Wilson (1999) The use of storytelling to understand women's desires to tan: lessons from the field, *Professional Geographer*, 51:2, 296-306.

Gesler, Wilbert (2003) *Healing Places* Lanham MD: Rowman and Littlefield.

Grimshaw, Jean (1999) Working out with Merleau-Ponty, in Jane Arthurs and Jean Grimshaw (eds) *Women's Bodies: Discipline and Transgression*, London and New York: Cassell, pp. 91-116.

Hertz, Rosanna (ed.) (1997) *Reflexivity and Voice.* London: Sage.

Hochschild, Arlie (1983) *The Managed Heart* Berkeley: University of California Press.

Hubbard, Phil, (2000) Desire/disgust: mapping the moral contours of heterosexuality *Progress in Human Geography* 24: 191-217.

Johnston, Lynda (1996) Flexing femininity: female body-builders refiguring the body *Gender, Place and Culture* 3: 327-340.

Kearns, Robin and Brendan Gleeson (2001) Remoralizing landscapes of care *Environment and Planning D: Society and Space* 19: 61-80.

Kobayashi, Audrey and Linda Peake (2000) Racism out of place: thoughts on whiteness and an anti-racist geography in the new millennium, *Annals of the Association of American Geographers* 90:2, pp. 392-403.

Longhurst, Robyn (2001) *Bodies: Exploring Fluid Boundaries.* London and New York: Routledge.

Mathee, Deidre (2004) Towards an emotional geography of eating practices: an exploration of the food rituals of women of colour working on farms in the Western Cape, *Gender, Place and Culture*, 11:3, 437-443.

McCormack, Derek (2003) An event of geographical ethics in spaces of affect, *Transactions of the Institute of British Geographers* 28: 488-507.

McKittrick, Katherine (2000) 'Black and 'cause I'm black I'm blue': transverse racial geographies in Toni Morrison's *The Bluest Eye, Gender, Place and Culture*, 7:2, pp. 125-142.

Mehta, Anna and Liz Bondi (1999) Embodied discourse: on gender and fear of violence, *Gender, Place and Culture* 6: 67-84.

Mestroviç, Stjephan G. (1997) *Postemotional Society.* London: Sage.

Milligan, Christine (1999) Without these walls: a geography of mental ill health in a rural environment, in Ruth Butler and Hester Parr (eds) *Mind and Body Spaces: Geographies of Illness, Impairment and Disability*. London and New York: Routledge, pp. 221-239.

Moss, Pamela (1999) Autobiographical notes on chronic illness, in Ruth Butler and Hester Parr (eds) *Mind and Body Spaces: Geographies of Illness, Impairment and Disability*. London and New York: Routledge, pp. 155-166.

Moss, Pamela, John Eyles, Isabel Dyck and Damaris Rose (1993) Focus: Feminism as Method, *The Canadian Geographer*, 37: 48-61.

Myslik, Wayne (1996) Renegotiating the social/sexual identities of places. Gay communities as safe havens or sites of resistance, in Nancy Duncan (ed.) *BodySpace: Destabilizing Geographies of Gender and Sexuality.* London and New York: Routledge, pp. 156-169.

Namaste, Ki (1996) Genderbashing: sexuality, gender and the regulation of public space *Environment and Planning D: Society and Space* 14: 221-240.

Nash, Catherine (2000) Performativity in practice: some recent work in cultural geography *Progress in Human Geography* 24:4, 653-664.

Parr Hester (2000) Interpreting the 'hidden social geographies' of mental health: a selective discussion of inclusion and exclusion in semi-institutional places. *Health and Place*, 6: 225-238.

Parr, Hester (1999) Delusional geographies: the experiential worlds of people during madness and illness, *Environment and Planning D: Society and Space* 17: 673-690.

Parr, Hester (forthcoming) Emotional geographies, in Paul Cloke, Philip Crang and Mark Goodwin (eds) *Introducing Human Geography*. London: Arnold.

Probyn, Elspeth (2003) The spatial impeative of subjectivity, in Kay Anderson, Mona Domosh, Steve Pile and Nigel Thrift (eds) *Handbook of Cultural Geography.* London: Sage, pp. 290-299.

Rose, Gillian (1997) Situating knowledges: positionality, reflexivities and other tactics *Progress in Human Geography* 21: 305-320.

Segrott, Jeremy and Marcus Doel (2004) Disturbing geography: obsessive-compulsive disorder as spatial practice, *Social and Cultural Geography*, 5:4, 597-615.

Sibley, David (1995) *Geographies of Exclusion.* London and New York: Routledge.

Smith, Mick (2001) *An Ethics of Place: Radical Ecology, Postmodernity and Social Theory.* New York: State University of New York Press.

Thien, Deborah (2004) Love's travels and traces: the impossible politics of Luce Irigaray, in Jo Sharp, Kath Browne and Deborah Thien (eds) *Women and Geography Study Group: Gender and Geography Reconsidered*, pp. 43-48.

Thrift, Nigel (2004) Intensities of feeling: towards a spatial politics of affect, *Geografiska Annaler* 86B: 57-78.

Thrift, Nigel and John-David Dewsbury (2000) Dead geographies and how to make them live again *Environment and Planning D: Society and Space*, 18: 411-432.

Valentine, Gill (1993a) (Hetero)sexing space: lesbian perceptions and experiences of everyday spaces, *Environment and Planning D: Society and Space*, 11: 395-413.

Valentine, Gill (1996) (Re)negotiating the 'heterosexual street': lesbian productions of space, in Nancy Duncan (ed.) *BodySpace: Destabilizing Geographies of Gender and Sexuality*. London and New York: Routledge, pp. 146-155.

Wood, Nichola (2002) 'Once more with feeling': putting emotion into geographies of music, in Liz Bondi, Hannah Avis, Amanda Bingley, Joyce Davidson, Rosaleen Duffy, Victoria Ingrid Einagel, Anja-Maaike Green, Lynda Johnston, Sue Lilley, Carina Listerborn, Mona Marshy, Shonagh McEwan, Niamh O'Connor, Gillian Rose, Bella Vivat and Nichola Wood, *Subjectivities, Knowledges, and Feminist Geographies*. Lanham MD: Rowman and Littlefield, pp. 57-71.

SECTION ONE
LOCATING EMOTION

Chapter 2

Placing the Dying Body: Emotional, Situational and Embodied Factors in Preferences for Place of Final Care and Death in Cancer

Sara M. Morris and Carol Thomas

Introduction

The location of terminal care and death for cancer patients is a highly emotive subject, fuelled by general taboos about untimely death, but also by anxieties about the particular degenerative processes of advanced cancer, which often leave the body 'unbounded' and out of control. Hospices and the new medical discipline of palliative care grew up around a desire to manage such problems, to reduce suffering and to locate the activities of dying and death in a 'safe' public space (Clark and Seymour 1999). A long-term trend away from home-based and towards institutionally located deaths has been taking place in the UK and other advanced industrial societies. Between 1967 and 1987, the proportion of cancer deaths in hospitals in England increased from 45 per cent to 50 per cent, and from 5 per cent to 18 per cent in hospices and other institutions (Cartwright 1991). The increase in hospice-based deaths is testimony to the rapid expansion and diversification of hospice services in the UK in the years following the opening of the landmark St Christopher's Hospice in 1967 (Clark and Seymour 1999). In 2000, the proportion of cancer deaths in hospital in the UK stood at 55.5 per cent, with 16.5 per cent occurring in hospice, 23 per cent at home, and 5 per cent in other settings, principally nursing and residential homes (Ellershaw and Ward 2003). Service provision varies across the country and this has been identified as one factor in place of death (Grande et al. 1998). Additionally, several studies have suggested a link between socioeconomic factors and place of death, with people from more deprived areas, or of lower social class, less likely to die in their own homes (Higginson et al. 1999).

UK health policies are currently replete with statements that services must reflect users' preferences and accommodate choice, and several studies have asked what preferences people have for place of final care and death, with results suggesting that most people want to die at home. Higginson and Sen-Gupta (2000) conducted a qualitative systematic literature review of patient preferences for place of care in advanced cancer, identifying 18 studies, half undertaken in the UK. Although indicating that half the studies were flawed in design or reporting, they conclude that home care is the most common preference for over 50 per cent of patients, with inpatient hospice care as second preference for those with advanced cancer. Higginson and Sen-Gupta (2000) also noted evidence that preference for *care* at home tended to be higher than for *death* at home, and that preferences tended to shift away from home and toward institutional care as death approached. These figures hint that, in advanced cancer, the increasing reality of bodily changes may prompt people to re-evaluate the simple proposition that 'home is best'. Other strands of literature indicate that choice in health care is influenced by emotional matters. Lupton (1996:166), for example, has argued that 'physical dependency goes beyond conscious rationality to a situation of semi-conscious or unconscious needs, desires and emotional states'. While Leich (2000: 209), writing about hospice at home in the USA, has indicated that the 'prevailing self-selection model' for choice in terminal care fails to acknowledge the 'complex, socially mediated processes' that surround these choices. Furthermore, she argues that this may be a contested process, and health professionals may influence care decisions around the location and type of final care. 'Choice' at the end of life, we believe, is not merely a matter of individual preference, but relates to socioeconomic status, service provision, cultural discourses, and emotional and relational factors; it is infused with the political and emotional geographies of care (Brown 2003).

It has been argued that the 'modern way of dying' from cancer is based on awareness, self-determination and heroism, thus linking with, and reflecting, modern pre-occupations with identity and consumerism (Seale 1998; Field 1996). The ideals of patient autonomy and choice are seen as central to the 'good death' (Leich 2000; Samarel 1991; also compare discussion in Hepworth, this volume). Increasingly placed within the concept of the 'good death', community and home are emphasised as the best places in which to end one's life. Clark and Seymour (1999) argue that current interest in the availability of home deaths confirms the strength of this notion. Institutionalised death is widely seen as a second best to the more 'natural' death at home. Yet within this seemingly simple proposition lies a dilemma: how can the Western values of autonomy be sustained where the 'natural' processes of dying may be 'dirty', and the 'unbounded body' may threaten the ability to maintain identity? Lawton (1998: 121), in her observational study of a contemporary UK hospice, suggests that the policy shift toward home death means that the hospice has become a 'no place', a place where the most extreme 'processes of bodily deformation and decay are sequestered', thus serving the needs of Western culture to maintain the ideals of personhood as sanitised and bounded. She argues that there is now a split between the type of (clean) dying that can be managed in the community, and the type of (dirty) death – accompanied by loss of self/body control – that is more likely to be located within the physical walls of an institution. People with advanced cancer find themselves placed within competing discourses; they are exhorted to exert the autonomy and choice prized in Western culture, while at the same time they are subject to processes of bodily decay that threaten their ability to maintain such a self-contained and self-determining 'self'.

Our study of preference for place of death among cancer patients and their informal carers took place against this background (Thomas, Morris and Clark 2004). Palliative care research has tended to focus on a simple stated preference, and there has been little specific exploration of how people with cancer and their carers actually negotiate this issue. It was certainly the case that home was the single most preferred location of death amongst our sample, but it was closely followed by a hospice preference and by an equal preference for either 'home *or* hospice'.[1] In this chapter we question the 'home is best' mantra, and wish rather to address the 'particular, relational, concrete and contextually embedded' (Kelly 2003: 2278) factors that relate to the question of where people would like to die. In this we focus particularly on 'home' as an emotional landscape reflecting identity and loss. Using extracts from patients and informal carers, we describe how choosing the 'right' place for death prompts discussion of normality and safety. We suggest that place of death it is not merely a matter of individual choice, but rather embedded in pre-existing relationships with place and other people. In this situation feelings about 'home' are often altered, and the divisions between public and private spaces may be blurred.

The Research Project

The data on which this chapter is based form part of a study undertaken in the North West of England in 2000-2002. The study had four key questions:

1   Where do cancer patients die in this area – at home, in hospice, in hospital, or elsewhere?
2   Why do they die in these locations?
3   What are patient and carer preferences for place of death?
4   Does the place of death match patient and carer preferences?

This chapter focuses on question 3, which comprised a series of interviews with cancer patients near the end of life and their carers. We aimed to obtain two face-to-face interviews with patients, and two subsequent short telephone updates. The warnings given in the palliative care research literature about the difficulties of undertaking research with dying patients proved to be well-founded (Rinck *et al.* 1997; Vigano *et al.* 2000; Christakis and Lamont 2000). Some patients became too ill to commence or continue with the series of interviews, or died earlier than expected: of the 41 patients who commenced the interview series, only 17 were able to see it through to completion. Twenty-four (59 per cent) of the participants were female and 17 (41 per cent) were male. Their average age was 67 years (range: 41-88). There was no resident carer in 17 cases. A variety of tumour types were represented, with larger proportions for the more common cancers. All patients and carers were from a White ethnic group (reflecting the composition of the population in the locality).

The sensitive nature of the subject matter and the physical status of participants meant that we conducted flexible interviews, which were conversational in style, although covering a range of pre-defined themes. The interviews explored these themes: quality of life, levels of pain and other symptoms, feelings about current care arrangements, and whether there were plans for future care. In addition, we prefaced the interview with a request for a narrative history of the cancer journey thus far. We also tracked care locations/patterns and the reasons for any changes (for example, hospital admission), and determined, through discussion, preferences for place of final care (a) given the existing situation and (b) if circumstances were 'ideal'. Preference for place of death was thus elicited through sensitive questioning on a range of related issues, and could be more or less direct depending on the ease with which the patient could address such matters explicitly. Carers were recruited into the study via our contact with patients. It was felt important that the patient should identify their main carer during the first interview. If the carer was co-resident s/he was usually asked for a (later) interview at the time of the first patient visit. In practice, the carer was sometimes present during the first interview with the patient (10 interviews). This was especially true where an older couple were involved. Our aim was to interview the carer once while the patient was alive and once after bereavement. In four instances the first interview with the carer had not taken place, although they had been asked and were willing. This was usually due to a sudden worsening of the patient's condition. In these cases the carer provided a bereavement interview only. Twenty-four co-resident carers were identified, and 18 agreed to take part. The carer sample consisted of more women (72 per cent) than men. The carers' average age was lower than that of patients at 61 years (range: 33-82). The majority of the carers (n=14) were in a spousal relationship to the patient, and sixteen had been living with the patient prior to the diagnosis.

Where to Die? The Place of Emotion

It might be thought that eliciting preferences for place of death, while sensitive, is a matter of uncovering something that exists as a 'pure' type, a clearly definable entity, that can be categorised. The research endeavour of identifying patient preferences is presented by many authors as a fairly straightforward procedure: once patients have spoken on the matter, boxes on research instruments can be ticked: home, hospice, hospital, other. From the start, we were somewhat sceptical about what constituted 'a preference'. Put another way, much of the research on preference for place of death has been undertaken within a positivistic philosophical framework (Higginson *et al.* 2000), and we were coming at the issues more phenomenologically, using qualitative research methods.

We wanted to ask dying people where they would like their final care to take place and we asked this question in several ways, one of which was a question about what they would want 'ideally'. It became apparent very soon that this was often a difficult question to answer in the abstract. Our participants nearly always located their responses in the detail of their personal circumstances, and preferences were hedged with provisos and uncertainties and speculations. Preferences were embedded in their personal contexts of symptoms, bodily matters, the nature of relationships with the informal carer/carers, and knowledge of and relationships with health services and health personnel. Preference for place of final care and death was not a concept people in our sample could discuss as a matter of simple and definitive choice, but was highly contingent. Expressed preferences were strongly coloured by emotional and social landscapes. It was highly mutable, and thus put into question the apparently clear-cut nature of many quantitative reports.

Table 2.1   Preference for place of death over time and actual place of death

Place of death preference	Preference at *start* of interview series	Preference at *end* of interview series	Actual place of death
Not decided	12	8	-
Home	10	6	8
Hospice	8	6	17
Home or hospice	8	3	-
No preference	2	1	-
Nursing home	1	0	1
Hospital	0	0	7
Not available	0	17*	8[‡]
Total	41	41	41

* Data was not available from second interviews with 17 of our sample, due to their death or worsening symptoms.
[‡] 8 of the patients were still alive at the end of the study period and it was unclear whether one patient had died at home or in a hospice.

Table 2.1 demonstrates our attempt to classify responses into preferences. The table shows that not only did expressed preference shift over time, but also the range of stated preferences as constituted by our sample were wider than merely home, hospice, hospital or nursing home, and in fact were often more akin to states of mind than actual places. Although several of our sample were very definite in stating a preference (and sometimes this preference was a clear '*no* preference'), the majority were ambivalent.

The following sections examine factors related to home as it was constituted in the data from our study.

Home

As we have noted, death at home is held as a 'gold standard' among many palliative care commentators (Stajduhar and Davies 1998). However, home is often viewed as a homogeneous and unproblematic location. In the palliative care literature home care is usually configured as cost-effective, more inclusive of patients and their family, and contributing to the overall quality of people's lives. In support of the latter point death at home is seen as facilitating a sense of normality, security and personal control (Stajduhar and Davies 1998). Yet home can be a place of struggle and physical, emotional and organisational labour (Williams 2002). Several studies have suggested that families caring for dying people at home frequently experience increased stress and pressure, even where there is support from community services (Cartwright and Seale 1990; Stajduhar and Davies 1998). Home may be a site of shared symbolic meaning (Relph 1976), but that meaning may not always conform to the idealised home that succours quality of life.

Our data provided examples that challenged some of these constructions of home as necessarily the best place to die. Williams (2002) suggests that home care impacts on perceptions of the meaning of 'home' and this certainly seemed to be the case for some of our respondents.

Sense of Normality

While clearly home is a more familiar environment than an institution, serious disease may render it noticeably less 'normal' as patients contrast their present selves with past selves:

> Female patient: It gets me so frustrated, I could scream, because when you've always been used to dashing around and doing everything yourself, it comes terribly hard after 75 years to not be able to do it. [age 79, lives alone]

Clearly, advanced disease brings changes to people's lives, and managing loss has been identified as a particular concern (Watson *et al.* 1988). Loss of the previously healthy 'self' and issues of dependency were constructed as 'abnormal' by many of our sample. In this situation it does not follow that the physical location alone will ameliorate loss. Williams (2002) in discussing 'therapeutic landscapes' as a framework for examining home space and health emphasises that these landscapes vary for individuals and that they vary over time. For people living alone, any choice to remain at home in advanced illness was almost certain to involve people coming into their home space and thus changing it:

> Female Patient: [a friend of mine] she's, 'No I'm not having anybody in my house', etc. I said, 'Don't be silly'. But I understand. To have somebody completely strange come into your house. But I think if I got to that stage, most probably I wouldn't mind. [age 84, lives alone]

Bodily changes could also affect the physical arrangement of the home environment in emotional ways. The normal routine of going up to the bathroom became a matter of dread for one male patient:

Male patient: I used to dread [going to the bathroom] in my old house. I had a set of stairs that I used to call the north face of the Eiger, and on one occasion I climbed up the stairs and it took me 40 minutes to recover, and I was going up on my hands and knees. [age 57, lives with friend]

For a carer the prospect of the patient's death occurring in the house was threatening to her use of her home:

Female carer: The Macmillan nurse was saying last week, 'How do you feel about somebody dying in your house?' I said, 'I'm not too happy about it, and I certainly won't be going in that room'. [sister, age 58]

To assume that home provides 'normality' ignores the complexity of emotional and social factors of place, and the identity changes incurred during the progression of cancer. As Cartier (2003: 2292) notes, 'changes in the body, and changes in social conditions [...] hold the potential to alter human relation to space'.

Security

People who lived alone often mentioned that home was not necessarily a 'safe' place any more. Regarding discharge from the hospice, a young woman said:

Female patient: I live on my own and I feel quite vulnerable when I'm at home. I guess that's the major part of it, as you suddenly change from having nursing on hand day and night to flying solo, and it's a big transition. [age 44, lives alone]

The assumption that home offers a secure location can be challenged by the circumstances of advanced cancer, particularly for those living alone. Even those who had a co-resident carer posited 'safety' concerns related to the home, generally concerning speed of access to help in a crisis. One couple had particular difficulties with out-of-hours care several times, where relief for extreme physical and mental symptoms was very slow in coming. The carer described her feelings thus:

Female carer: It's very frightening to think that if this happens again, how will I be certain that it isn't another five hours [before anyone comes]? There is this element of risk, which I'm now very aware of. [...] I have felt very frightened and insecure. What happens if-? And it's night time, you know. [wife, age 63]

Safety is here associated with accessible 'symptom control'. The patient had strongly stated his preference for death at home and, as his difficult dying process progressed, his wife loyally defended his right to self-determination, despite its traumatic effect on her own sense of security. For her, however, the home was rendered an unsafe site, at least during the 'out-of-hours' periods.

Personal Control

The idea of greater freedom and control in the home setting is usually illustrated as offering the opportunity for patients and carers to follow their own routines, in contrast with institutional regimens. However, community services, which are invariably needed when the patient lives alone, or to support a carer, can entail their own impositions. One woman in our sample had refused help from formal carers because they had offered to come at 7am to get her out of bed and this was completely at odds with her normal routines. Another man who lived alone refused home help because:

> Male patient: I don't want anyone rummaging through my drawers. I mean, as long as I can manage. I can keep my house tidy and do my washing and all that sort of thing, as long as I can manage, that's how I'll carry on. [age 88, lives alone]

Sometimes the offer of support at home was perceived differently by patient and carer, and could have led to conflict. One woman described how a care package had been arranged for her brother against his wishes, but in line with hers:

> Female carer: The social services had come down and sorted it all out. 'We'll have this package in place for when he's discharged.' He wasn't happy about it: 'I don't want somebody coming in and getting me up for me breakfast and helping me to dress and wash'. I said, 'Well, we'll jump that hurdle when we come to it. Get into the hospice and get your rest first'. That's how you had to deal with him, but then it didn't come off you see, he didn't come home, but the package was being put in place ready for him which would have been a godsend. [sister, age 66]

For people in our sample, then, personal control was not automatically sustained in the home location when formal home care was instigated. In addition there were indications that the people around the patient and the relationships involved were important factors in feelings about location. In the next section we explore the impact of relationships with informal carers on preference for place of care.

Relationships with Informal Carers

As well as being a physical location, home is also the crucial setting through which basic patterns of social relationships are constituted and reproduced (Walmsley and Lewis 1993). One of the main predictive factors for people dying at home is the presence of a willing and able carer (Cantwell *et al.* 2000). Clark and Seymour (1999: 33) remind us that '"caring" has been identified as both the range of activities involved in giving personal aid and support to someone in need, *and* the emotions associated with an interpersonal relationship'.

Care activities are interpreted as evidence of affective ties between people, especially where those involved are related by kinship. One carer, despite suffering from ME, was determined to keep her husband at home and in her care until the end, which she did. She evoked the power of love as enabling her to manage him, even when he was needing a great deal of physical assistance:

> Female carer: I could – I mean he weighed so little, you know.
> Interviewer: But you're not very big either are you?
> Carer: No.
> Interviewer: So lifting him?
> Carer: But I think sheer love makes you do it. Sheer love gives you the strength to do it. [wife, age 70]

However, different scenarios were also possible; although most spouses were described as very supportive, one woman described her husband as unsupportive, unwilling even to take her to the hospital for appointments. Her sense of betrayal and need to be cared for affected her preference for place of death:

> Female patient: I mean, well, what are we going to do at the end? So I just think, 'Well, I hope I end up in the hospice'. At least they look after you, you know. I mean you don't want to, but I think, 'Well, I won't get any support here'. [age 61, lives with husband]

The association between home death and informal carers is one of the 'most consistent themes in the literature' (Brown and Colton 2001: 801). In our interviews personal social setting and informal support networks were always centrally placed in patients' talk about preference. End of life care is embedded in pre-existing social circumstances, and consideration of how these might be altered and how carers might cope are questions raised by the situation of terminal cancer. Even those who lived alone were concerned about any family they had, and sometimes about friends.

It has been noted that family responsibilities are fluid and variable, rather than fixed, and that reciprocity and dependence will have a range of meanings depending on negotiations within specific interpersonal relationships. Clark and Seymour (1999: 39) indicate that 'Problems of role conflict between the demands of caring, employment and other pressures may be differentially interpreted and resolved' in the situation of terminal care. Concerns about 'being a burden', and the appropriateness of the relationship with the carer/s, were highly relevant to preferences. For example, one woman who had nursed her husband with cancer, and thought that this was right and proper, nevertheless knew what it had been like for her and wanted to protect her children from the distress:

> Female patient: If I was really sick, I mean *sick* sick, I wouldn't want my kids to look after me. I think I'd sooner go somewhere to be looked after. I hope it doesn't come to it. Well, something's got to happen some time, but because they were so good with their dad and they went through so much, I wouldn't want that again. [...] If I could spare them that, well I would. I mean I had it to do, as [he was] my husband, but it was too much for them and I wouldn't want it again so we'll have to see. [age 70, lives alone]

Familial relationships also seemed to have a bearing on preference for place of care in the event of physical dependency and the need for intimate nursing. Put generally, spouses were seen as appropriate carers, but other familial relationships were less favoured by many. Sometimes feelings of embarrassment were mentioned by respondents in relation to same sex relationships (such as mother/daughter), but were much more frequently invoked where the relationship was not spousal and the patient and carer were of opposite sexes. We had two cases of mother/son co-residence, where each mother valued their son's help with practical matters and their company, but neither felt it was right for their sons to nurse them if they became bedridden.

> Female Patient: I don't honestly know [what would happen]. I suppose if I came to be bedfast-. He [my son] couldn't do it, could he? [age 75, lives with son]

This concern was also shared in sibling relationships in our sample, where the patient was the brother and the carer the sister. In both recorded cases neither patients nor carers felt that their relationship should involve intimate nursing. Here is one pairing's account:

> Female carer: Now, this blip [incontinence] that he had last week happened twice and he was beside himself, and I just made light of it. I said, 'Oh forget it, it's nothing, don't worry about it'. And I thought, 'I can't cope with this'. I'm not a nurse. It's not your child that's been sick. This is a man and he's full of cancer. And you somehow get this smell in your house that you … Now that's what hospitals are for. [sister, age 58]

> Male patient: What I'd like to do basically is go to bed one night, die and not wake up. It wouldn't bother me if it were tonight you know, but what I'm saying is that I don't want to be getting incontinent and stuff like that and having my sister, well putting her through it. I know it's the same for the nurses, but they're used to it, aren't they? [age 57, lives alone]

Anxieties about the particular degenerative processes of advanced cancer leaving the body 'unbounded' were common in our sample. As the previous examples suggest, these worries were closely entwined with considerations of relationships with informal carers. Home – in these instances of strong feelings about not imposing bodily degeneration on those close to you – was a place to avoid. Institutional locations and the care of professionals were often the preferred means of managing emotions such as embarrassment and disgust and distress.

Typically, in the social science literature, home has been discussed as a 'private' space and institutions as more 'public' (Cornwell 1984; Twigg 1999). But in these instances distinctions between private and public spaces were blurred by emotional and social considerations. Twigg (1999) argues that a re-ordering of the public/private division occurs when people require intimate care. In situations of high bodily dependency 'home' does not always sustain the 'privacy ethic'. Hospices offer a site where the drawing of boundaries around bodies happens at a literal, physical level, a place where it is possible to maintain a separation between the hygienic life world and the disintegration of the flesh (Lawton 2000), and where containment of death imposes order and meaning and the threat to the remaining social body diminishes (Mellor and Shilling 1993). Lawton (2000) argues that hospice is a 'no place' where death is overt, but 'dirty' processes of dying are veiled, glossed over, as symptoms requiring control, thus reinforcing contemporary western cultural values of individualism and self-containment. Many of our participants expressed these prevailing values, and indicated that particular situations warranted the re-ordering of public and private. However, stating a preference for removing the unbounded body from habitual social settings, and/or into the care of professionals, was highly contingent for both patients and carers, and depended to a large extent on judgements about the appropriateness of the social relationship with the carer.

Concluding Remarks

Although only briefly sketched out here, these kinds of findings lead us to question the assumption that stating a preference for home death is a simple matter of individual choice as health policy appears to assume (Department of Health 2000). The meanings of home and the relationships embedded in this place strongly affected our sample's talk about preference. In addition, people were more concerned with the period of dependency leading up to death than the moment of death. It was not so much the 'where' of dying, but the 'how' and 'with whom' that seemed to matter.

For our sample, privacy and dignity were constructed as components of relationships rather than properties of a place, and distinctions between public and private spaces were blurred by emotional and social considerations. It did not automatically follow that the home situation would minimise distress. In some circumstances it might threaten to exacerbate it. Emotional concerns about safety, intrusion, dependency, loss and dignity were not simply solved by the home location. Our findings suggest that a range of flexible services is needed, if cancer patients and their carers are to have their complex and shifting preferences met.

Acknowledgements

We are grateful to the patients and carers who agreed to participate in the research at such a difficult time in their lives. The health professionals who have assisted the project in a variety of ways are also owed many thanks. We are grateful to the Research and Development Department, NHS Directorate of Health and Social Care North for generously funding this research.

Note

1   It should be noted that in the context of the study reported here, the term 'hospice' refers to a physical building in which people are cared for and may die, rather than a philosophy of care than can be delivered in home and other settings. The area has two hospices of differing sizes, but in combination they represent a provision of specialist palliative care beds for the district that is above the per capita national average.

References

Brown M (2003) Hospice and the spatial paradoxes of terminal care. *Environment and Planning A* 35: 833-851.

Brown M and Colton T (2001) Dying epistemologies: an analysis of home death and its critique. *Environment and Planning A* 33: 799-821.

Cantwell P, Turco S, Brenneis C, Hanson J, Neumann CM and Bruera E (2000) Predictors of home death in palliative care patients. *Journal of Palliative Care* 16(1): 23-8.

Cartier C (2003) From home to hospital and back again: economic restructuring, end of life, and the gendered problems of place-switching health services. *Social Science and Medicine* 56: 2289-2301.

Cartwright A (1991) Changes in life and care in the year before death 1969-1987. *Journal of Public Health Medicine* 13(2): 81-87.

Cartwright A and Seale C (1990) *The natural history of a survey: an account of the methodological issues encountered in a study of life before death.* London: King's Fund.

Christakis NA and Lamont EB (2000) Extent and determinants of error in doctors' prognoses in terminally ill patients: prospective cohort study. *British Medical Journal* 320: 469-473.

Clark D and Seymour J (1999) *Reflections on palliative care.* Buckingham: Open University Press.

Cornwell J (1984) *Hard-earned lives.* London: Tavistock.

Cuba L and Hummon D (1993) A place to call home: identification with dwelling, community and region. *The Sociological Quarterly* 34: 111-131.

Department of Health (2000) *The new NHS Cancer Plan.* London: Department of Health.

Ellershaw J and Ward C (2003) Care of the dying patient: the last hours or days of life. *British Medical Journal* 326: 30-34.

Field D (1996) Awareness and modern dying. *Mortality* 1: 255-66.

Goddard MK (1993) The importance of assessing the effectiveness of care: the case of hospices. *Journal of Social Policy* 22: 1-17.

Grande GE, Addington-Hall JM and Todd CJ (1998) Place of death and access to home care services: are certain patient groups at a disadvantage? *Social Science and Medicine* 47: 565-79.

Higginson IJ, Jarman B, Astin P and Dolan S (1999) Do social factors affect where patients die: an analysis of 10 years of cancer deaths in England, *Journal of Public Health Medicine* 21: 22-28.

Higginson IJ, Astin P and Dolan S (1998) Where do cancer patients die? Ten-year trends in the place of death of cancer patients in England, *Palliative Medicine*, 12: 353-63.

Higginson IJ and Sen-Gupta GJA (2000) Place of care in advanced cancer: a qualitative systematic review of patient preference. Journal of Palliative Medicine 3(3): 287-300.

Hospice Information (2001) Minimum Data Sets national survey 2000/01. http://www.hospiceinformation.co.uk/Facts/facfig1.asp

Kelly SE (2003) Bioethics and rural health: theorising place, space, and subjects. *Social Science and Medicine* 56: 2277-2288.

Lawton J (2000) *The dying process: patients' experiences of palliative care.* London: Routledge.

Lawton J (1998) Contemporary hospice care: the sequestration of the unbounded body and 'dirty dying'. *Sociology of Health and Illness* 20(2): 121-143.

Leich J (2000) Preventing hospitalisation: home hospice nurses, caregivers, and shifting notions of the good death. pp. 207-228 in Kroenfeld JJ (ed.) *Health, illness and the use of care: the impact of social factors.* Volume 18. JAI London: Elsevier Science.

Lupton D (1996) 'Your life in their hands': trust in the medical encounter. pp. 157-172 in V James and J Gabe (eds) *Health and the sociology of the emotions*. Oxford: Blackwell.

Mellor PA and Shilling C (1993) Modernity, self-identity and the sequestration of death. *Sociology* 27(3): 411-431.

Relph E (1976) *Place and placelessness*. London: Pion.

Rinck GC, vanden Bos GAM, Kleijnen J, deHaes HJCM, Schade E, Veenhof CHN (1997) Methodological issues in effectiveness research on palliative cancer care: a systematic review. *Journal of Clinical Oncology* 15(4): 1697-1707.

Samarel N (1991) *Caring for life and death*. New York: Hemisphere.

Seale C (1998) *Constructing death*. Cambridge: Cambridge University Press.

Stajduhar KI and Davies B (1998) Death at home: challenges for families and directions for the future. *Journal of Palliative Care* 143: 8-14.

Thomas C, Morris SM, Clark D (2004) Place of death: preferences among cancer patients and their carers. *Social Science and Medicine* 58(12): 2431-2444.

Twigg, J (1999) The spatial ordering of care: public and private in bathing support at home. *Sociology of Health and Illness* 21(4): 381-400.

Vigano A, Dorgan M, Buckingham J, Bruera E and Suarez-Almazor ME (2000) Survival prediction in terminal cancer patients: a systematic review of the medical literature. *Palliative Medicine* 14: 363-374.

Walmsley DJ and Lewis GJ (1993) *People and environment: behavioural approaches in human geography*. London: Longman Scientific and Technical.

Watson M, Greer S and Thomas C (eds) (1988) *Psychosocial Oncology*. Oxford: Pergamon Press.

Williams A (2002) Changing geographies of care: employing the concept of therapeutic landscapes as a framework in examining home space. *Social Science and Medicine* 55: 141-154.

Chapter 3

'Mourning the Loss' or 'No Regrets': Exploring Women's Emotional Responses to Hysterectomy

Marion Collis

Introduction

Drawing on interviews with 20 Australian women, this chapter explores the commonalities and differences in emotional responses to hysterectomy, that is, the surgical removal of the womb or uterus. While my focus is on direct experience as a way of knowing, I understand women's experiences of their bodies, and their responses to those experiences, to be socially and culturally produced (Wolff 1990). Thus, women's experiences of hysterectomy are not only informed by the dominant modes of thinking about the body in general, and the female body in particular, but also need to be understood within the context of women's socially ascribed identities and the ways in which these change across the life course.

The highly influential discourse of scientific medicine has resulted in the predominance of a very mechanistic understanding of the human body, in which parts can be removed, or even replaced, if they malfunction (Russell and Schofield 1986; Birke 1999; Lupton, 2003). In addition, women's bodies have traditionally been viewed as abnormal, or at least potentially disease-producing, since they are different from men's bodies, which are defined as the norm (Rowland 1988; Moscucci 1990). The result, according to Birke (1999: 160), is that 'woman [becomes] the add-on extra, the one with the removable uterus'. In fact, hysterectomy has been identified as 'the most common major surgical procedure undergone by women in many Western countries'[1] (Dennerstein et al. 1994: 311).

The medical/gynaecological literature has historically perceived women's emotional response to hysterectomy in essentially pathological terms, identifying high levels of psychiatric disturbance in women who had undergone hysterectomy. This has been attributed to the presumed psychic importance of the womb as a symbol of womanhood (Barglow et al. 1965; Hollender 1969). A review of the literature published between 1941 and 1970 entitled 'The crisis of hysterectomy' (Raphael 1972: 106) found evidence that the surgical removal of the womb 'may lead to a crisis situation in at least a percentage of women' resulting in depression and other psychological disorders. However, there was also a recognition by the author that the 'feelings of sadness and mourning' experienced by many women might be an adaptive and normal process (p. 110) and not necessarily a precursor of mental illness (see also Morris and Thomas, this volume). Drummond and Field (1984: 110) also noted the normality of an emotional reaction to the 'loss of part of the body and of valued body functions'.

In contrast to the earlier studies, studies since the 1970s have consistently failed to find evidence of a link between hysterectomy and increased levels of psychological/psychiatric illness (Carlson *et al.* 1993; Ryan 1997; Reid *et al.* 2000). Further, some research has actually identified a post-operative decrease in depressive symptoms following alleviation from the distressing physical symptoms of menstrual disorders (Alexander *et al.* 1996; Davies and Doyle 2002). While Ryan (1997) explains the contradictory results as largely due to methodological problems associated with the earlier studies, others have suggested that the changes over time may have more to do with the impact of second wave feminism and the changing role of women (Drummond and Field 1984; Cohen *et al.* 1989). According to this argument, since women's identity is no longer defined solely in terms of motherhood, there has been a reduction in the 'defeminising effects of hysterectomy' (Gould, 1986 as cited in Cohen *et al.* 1989) and therefore of the likelihood of serious psychological consequences.

To date, the vast majority of research on the physical and psychological aspects of the recovery from surgery as well as the longer-term outcomes of hysterectomy has been quantitative in nature and clinical in focus. These studies often take the form of randomised trials designed to compare the outcomes of different types of hysterectomy (Garry, *et al.* 2004) or hysterectomy and alternative procedures such as endometrial ablation (Alexander *et al.* 1996; Sulpher *et al.* 1996). While studies that focus on 'consumer satisfaction' (Schofield *et al.* 1991; McKenzie and Grant 2000; Nathorst-Boos *et al.* 1992) and 'quality of life' (Carlson *et al.* 1994; Rannestad *et al.* 2001; Davis and Doyle 2002) have begun to give more consideration to women's own views of their life after hysterectomy, including their emotional well-being, the research is still largely quantitative and provides little understanding of the biographical and social contexts behind women's responses. As Farquhar *et al.* (2002: 204) note in response to their findings that satisfaction with surgery and regret for the loss of fertility may co-exist, 'the question of satisfaction with the procedure may not be sufficient to detect the underlying value that a woman places on the procedure'.

Nursing research has tended to make more use of qualitative methods, including Kinnick and Leners (1995) who interviewed six women three months after surgery, and Fleming (2003: 576) who undertook a case study to 'explore the meaning of hysterectomy for one woman'. However, some qualitative studies have failed to provide more than a superficial analysis that adds disappointingly little to quantitative studies. For example, Linenberger and Cohen (2004) interviewed 58 women who had undergone hysterectomies for non-cancerous conditions. Despite reporting that two women felt 'sad at times' and two had 'a day of the blues' after their surgery (p. 353), there is no information about these women that would allow the reader to understand the meaning of these responses in the context of their everyday lives.

Thus, it is still largely true that 'women's lived experience of recovery from hysterectomy has not been well documented' (Linenberger and Cohen 2004: 350) and also that 'women's own evaluation of their personal experience needs to be investigated for accurate conclusions to be drawn about their reactions' (Reid *et al.* 2000: 4). Hopefully the research discussed in this chapter will provide a useful starting point for filling these gaps in knowledge.

The Research Project

As Roberts (1992: 5) notes in her Introduction to *Women's Health Matters* 'there remains a deep-seated suspicion (within the medical profession) of women's own accounts' of their reproductive and health concerns. When Brown and her colleagues (1994: 2) submitted a paper on post-natal depression to a prestigious medical journal, the editors rejected it on the grounds that the women's (subjective) accounts of their birth experiences were 'extremely dubious' since they could not be objectively validated. This research clearly rejects this position, sees women as 'knowledgeable actresses' (Sydie 1987: vii) and privileges the knowledge that comes from personal experience (see also Bondi, this volume).

The 20 women whose experiences form the basis of this chapter were recruited by placing notices about the research in Community Health Centres in the eastern suburbs of Melbourne and the Latrobe Valley region of country Victoria, Australia.[2] I sought to interview women under 50 years of age who had undergone a hysterectomy within the previous five years, and whose operations had been for menstrual disorders produced by benign conditions.[3]

The women all participated in in-depth tape-recorded interviews of between 30 and 90 minutes in length. As Bell (1988: 101) notes:

> ... in in-depth interviews, people spontaneously tell stories to tie together significant events and important relationships in their lives, and to 'make sense' of their experiences. Through linked stories, people explain how their experiences – and their interpretations of these experiences – have changed over time.

While a chronological framing was used to structure the interviews, the women's narratives themselves were more fluid and complex. The women often revisited key events in their menstrual and treatment histories later in the interviews as they remembered further details or as new insights came to them. Such reworking of their individual hysterectomy stories was frequently interwoven with significant events in their marital and childbearing biographies. All the women were, or had been married, and came from Anglo-Australian backgrounds. Their median age at the time of surgery was 38 years, with the youngest being aged just 29 years and the oldest 48 years.[4] The respondents all had at least one child, with most having two (8) or three (9). The main reasons for needing a hysterectomy were fibroids (7), endometriosis (3), prolapse (3) and heavy bleeding and/or menstrual pain (7).[5]

Exploring the Women's Stories

Framing the Narratives: Life Course, Loss and Emotional Support

Williams (2000: 51) defines the life course as a 'social clock' that 'guides our experience of events within the biographical context'. Medical discourse has played an important role in constructing women's life course around their reproductive cycle, suggesting that biographical transitions will bring with them changed understandings of femininity and what it means to be a woman (Abbott and Wallace 1998). In their study of early menopause, Singer and Hunter (1999: 2) identified the emotional distress caused by 'deviation from age norms', specifically the loss of reproductive capacity at the stage of the life course when femininity is seen to be closely associated with fertility. This suggests that the experience of hysterectomy and the meaning which a woman gives to that experience (how she understands and names the experience (Lupton 1998)) is likely to vary according to whether or not the loss of her reproductive capacity is experienced as a 'biographical disruption' (Bury 1982; Williams, 2000), that is, the extent to which it shatters a woman's taken-for-granted understanding of her body and her sense of herself as a woman. However, while the experience of hysterectomy is shaped by social and cultural understandings of particular stages in the life course, each woman's experience is also unique. As well as identifying commonalities amongst respondents' experiences it is also important to acknowledge their diversity since 'the nature of experience is different for each person' (Fleming, 2003: 576).

As the women told their diverse personal stories, their reflections often focussed on the conception of hysterectomy as 'loss' and what that loss had meant to them. While respondents talked about the loss of menstruation and the loss, or otherwise, of sexual desire and/or satisfaction, most often in response to a specific question, they were more likely to spontaneously mention the loss of childbearing. This suggests that the loss of childbearing had greater salience for them in the context of their post-hysterectomy lives. The loss of menstruation was mourned by only three of the (younger) women – Amy (35), Catherine (32) and Debbie (33). For Amy and Debbie grief was more than offset by feelings of relief. Only Catherine felt that her identity as a woman was affected by the loss of menstruation. The majority of the women were only too happy to be free of the pain, mess, embarrassment (actual or potential) and general debilitation they had often experienced for some years. Again, only Catherine felt that there had been a loss of sexual satisfaction since her surgery, while for most respondents things had either changed little or for the better, largely due to the relief of their menstrual symptoms.

Although the importance of childbearing and the salience of that loss were identified throughout the narratives, the focus of the loss varied. Respondents talked about the loss of their future/potential childbearing capacity, the loss of their wombs (as an organ valued for its reproductive capacity) and the loss of their femininity or 'womanhood' (as intimately connected to a woman's ability to have children). As Bernhard (1992: 180) noted, a key factor affecting a woman's experience of hysterectomy is 'the relationship with [her] husband or sexual partner and the support provided by that person'. The women's accounts of their need for emotional support and the extent to which this was provided by their partners is therefore integrated into the discussion in order to provide some insight into the wider emotional context of the women's own experiences.

Loss of Future Reproductive Capacity

Sarah (34) had put up with extreme pain and heavy bleeding for two years while trying to get pregnant with her first child. After the birth of her son things became so bad that she felt there was no real option but to have a hysterectomy; this was in spite of the fact that both she and her husband wanted another child. Sarah's hysterectomy thus represented 'a lost life-plan' (Singer and Hunter 1999: 67). After the operation her distress was intense:

> What happened to me? Oh God, I'd sit there [on the ward] for breakfast and I'd just put my head down and cry and cry and cry and cry ... I said: 'Why me, why? Did I ask for this?' That's what I felt ... nobody asks for those kinds of things. Nobody can tell you: 'Hey pull your socks up, you've got to accept it'.

Sarah likened her situation to that of 'losing a child through cot death or something' and described herself as 'grieving for at least six months' after the operation. While her need for emotional support during this time was high, she found that her husband's ability to support her was limited:

> He didn't know how to cope, because it was very heavy with my crying and that in the first few months ... And I thought: 'Well, he doesn't understand', he did to his own degree being a man.

In the end Sarah decided not only that her husband did not really understand, but also that it was unfair to continue burdening him with her feelings because 'he was there through it too' and she felt that he could not take any more. So she decided that she would try and do her crying alone, in effect putting her husband's needs before her own, regardless of the emotion work required.

While Sarah was the only woman who had wanted more children at the time of her hysterectomy, Amy had faced a similar situation when first presented with the possibility of a hysterectomy after the birth of her second child at the age of 31. She had very much wanted another child and explained how much emotion work it took trying to pretend otherwise:

> So, we [my husband and I] thought, right we'll be very happy with the two children that we've got, which we were. But it was, I always had to work hard to be very happy. I was, there was always something underlying, not so much for my husband but for me ...

Just two weeks before she was due to go into hospital to have the hysterectomy Amy discovered that she was pregnant again. She finally had the operation four years later when her third child was three years old. Although she described herself as someone who 'would always want children no matter how many she had', she felt that by then 'time had marched on' and she was beginning to relinquish such desires. In fact her husband had had a vasectomy after their third child was conceived in order to give her time to get used to the idea of not having any more children before she underwent the hysterectomy. Even so, Amy found the loss of her childbearing capacities hard to cope with, especially as she 'still had friends that were having children'.

Catherine was in a similar situation, having started and finished her childbearing ahead of all her contemporaries. However, her grief was also tied up with the loss of the possibility of replacing a child who died:

> My youngest child's only just turned three anyway so I mean there's always this wavering down in the back of your mind that you might lose one or whatever and if you're still able to reproduce you can substitute a loss. [...] I could have changed my mind at any time, but I don't even have that to fall back on. Even my kids say: 'Why don't you have another baby?'.

Catherine noted that her husband had 'been a bit more tender [since her hysterectomy] ... both on the physical and emotional level; he's listening and trying to pacify me'. However, like Sarah's husband he had been unable to give her quite what she needed:

> he feels very um, unsure of what to say, because he doesn't know what the right thing is to say back. I mean, he wants me to feel good and reassured, but he doesn't know the way to do that.

Several other respondents who had completed their families also grieved for the loss of their wombs and the ability to become pregnant again. Dael (29), whose youngest child was also three, described a period of depression 'a bit like baby blues' that lasted for about three weeks. This had come as a bit of a surprise because she 'didn't want any more children' and her husband, like both Amy and Catherine's husbands, had had a vasectomy. As Dael reflected, the reason she had experienced a period of grief was 'probably 'cause everything was just final'. In contrast to Sarah and Catherine, Dael described her husband as 'very good' because he was able to help her through her grieving. However, she also noted that she was 'very fortunate' in this respect because there were other women 'out there' that did not have supportive partners.

Research on male carers has found that 'men are more likely than women to take an instrumental, task-orientated approach, reducing caregiving to a series of specific activities while remaining somewhat distant and detached' (Abel, 1990: 72). This, of course, suggests that women who are being cared for may lack the emotional support they need. This is exactly what happened in Jasmine's case. Like Dael, Jasmine (30) described her feelings after the hysterectomy as being similar to the depression she had experienced after giving birth, although 'perhaps not as intense'. Although she had not been planning to have any more children 'and had had her "tubes tied"', she found herself grieving for her lost womb 'and the ability to ever procreate'. Unlike Dael's partner, Jasmine's husband seemed to have little understanding of her need for emotional support. For him, 'having surgery was standard procedure, as fixing the problem'. So, when she tried to talk to him about how she felt 'he'd more or less turn around and say: "Take an extra tablet; it will quieten you down" or something'. Since her hysterectomy Jasmine's marital relationship has broken up and she has found a new partner whose 'sensitivity' she can 'really respond to'. However, her new partner does not have any children of his own, and Jasmine spoke of having 'dreams of another child' even though it was 'situation impossible'. She recalled that for a while this had made her feel 'a little inadequate' even though her partner was very accepting of the situation. Roeske (1978: 224) also noted that regret over hysterectomy might be linked to the importance for a woman who was not married, or who was divorced or widowed, to retain her childbearing capacity in order to 'have a child to please her future husband'.

In contrast to the above accounts from women who grieved for the loss of their future reproductive capacities, there were those who totally rejected any such suggestion:

> Oh no, that [losing my womb] didn't worry me, 'cause I didn't want any more kids. (Demon, age 42)

> No [I didn't grieve for its loss], not at all. I'd had my children and I mean I was 43 or 44 when I had the operation and I felt my childbearing days were over. (Lyn)

> ... grieving for the loss of the womb, that's never entered my head ... I've got more important thing to do with my life than start worrying about things like that. I'd decided that I'd have four children, and the youngest would have been before I was 30, and that was it. I mean I didn't want to have any more children after that. (Janet, age 41)

All three of the women quoted above were in their early forties at the time of their hysterectomies and the youngest child in each family was between 12 and 16 years old. In the way that their life courses had unfolded, each was long past thinking about having any more children. As Janet's comment indicates, her life had moved on from being closely bound up with having (and looking after) children, as it had been when they were young. Now she had other things she wanted to do with her life. While the need for emotional support was not as great for women like Janet, Lyn and Demon as it had been for the younger women, it was still important for them to feel that their partners' support was there if it was needed. It was not necessary for it to be constantly verbalised: it could be quite unobtrusive and the knowledge that their husbands were there for them was enough. Kinnick and Leners (1995: 151) in their study of the 'hysterectomy experience' also noted the importance respondents placed on having significant others who were 'there for me', that is, who showed concern and looked after their wellbeing.

However, even though she did not experience any grief over the loss of her reproductive potential, when Lynette (38) commented that having a hysterectomy would have been 'a different ball game altogether' if it had 'short-circuited having children', she was speaking for other women as well. Anne (39 with children aged 12, 17 and 19 at the time of her surgery), for example, said that she 'might have opted to ask more questions' of the gynaecologist when he told her that she needed a hysterectomy if she had 'been younger and had younger children'. Like Catherine, she would have wanted to retain the ability to replace a child if the unthinkable had happened and she had lost one of her children. Other respondents either commented that they would not have made the decision to have the operation if they had not completed their childbearing, or acknowledged that if 'it had been a matter of having to, then it would have been much more upsetting' (Susan, aged 44). Hallowell and Lawton's (2002) study of women at risk of hereditary ovarian cancer confirms the importance of women's reproductive potential and its impact on the decisions they make about the removal of their reproductive organs. The women who had chosen the option of annual screening stated that they would only have surgery (prophylactic removal of the ovaries) once they had finished childbearing, while those who had undergone surgery said that they had only agreed to it once 'they were certain that they had wanted no more children' (Hallowell and Lawton, 2002: 432).

Rosalie's story is a further illustration of how a woman's emotional response to the loss of childbearing is likely to change over her life course. When a hysterectomy had first been suggested to her at the age of 34 she totally rejected it, although she 'had no mental images of this representing my femininity or anything'. However, four years later she was happy to go ahead and there were 'no regrets'. She explained her changing response in the following way:

> ... [at the age of 34] the main thing was the ability to have kids. I realised as I say that is what completely threw me in '86. There was a chance then that the marriage, as I say, would have broken up and I wanted to have the facility, obviously subconsciously part of the marriage package is the ability to have kids for him. [...] It was, in analysis, it was my marriage value ... my chances of marrying again would be lessened a great deal if I didn't have the ability ... to have kids ... In '86 I wasn't working, the only sense of value I had was as a wife and mother. [...] whereas [at the age of 38] in '90 when I made the decision, I had been back in work and I now was getting back a sense of value of 'I am a person, I've got value' ... So in '86 the chance of remarriage, that I couldn't reproduce, what was there left? Whereas by '90 when I made the decision my own self-confidence, my personal self-confidence had completely come back, um I had value in the world.

In 1986 Rosalie did not want any more children, but could not come to terms with losing the ability to reproduce, because if her marriage should break up she felt that was all she had to offer a future partner. However, by 1990, when her two children were aged nine and eleven, she was back in the paid workforce, had found alternative sources of identity and self-worth as a woman and would not have wanted to 'start having babies again' whatever happened. Rosalie was also one of the women 'out there' who Dael talked about as not having their partners' support. Her husband had shown little interest in, or understanding of, the emotional aspects surrounding hysterectomy, in spite of her attempts to involve him in the decision-making. Although Rosalie did not experience any negative emotional responses after the surgery she commented that her husband had been surprised when it did not prove the panacea for all that was wrong with the marriage.

While Cathy (31) did not go through a grieving process following her hysterectomy, her reactions to other women's plans for sterilisation did suggest that there was still some level of regret at the loss of her childbearing abilities:

> ... though I know I did, sort of at the time and sort of for a period after that, if I heard of anyone that was having sort of sterilisation you know, I'd sort of think ... Oh, I know I sort of said to them: 'Oh, I'd really think about this, you know if contraception isn't a real problem with you, really, you know, don't throw it away'.

Like Sarah, Cathy had a young baby a few months old at the time of her hysterectomy. Unlike Sarah, this was Cathy and her husband's third child and quite possibly their last, even without the hysterectomy. In her interview Cathy noted that in the five years since her hysterectomy she had never been through the 'I wish I could have another baby'. While she was 'sort of waiting [for the depression] to come along' it never did, which she put down to the fact that she breastfed throughout her hospital stay and recovery.

Loss of a Valued Body Part

Like some of the younger women quoted above, Debbie also found herself grieving for her womb for 'about three to four weeks after the operation'. However, her feelings were not so much bound up with fantasising about its future reproductive possibilities as with its past reproductive abilities. Hysterectomy meant the loss of a valued body part that had given her five children, and which needed to be mourned just 'like it was a loss of a loved one'. She had even thought about giving it 'a proper burial':

> ... it was really funny, I'd made [a] comment to my husband before I had the operation, I'm going to ask for my uterus to bring it home and plant it in the back yard ... 'cause I really felt that it was, it had given me five lives, I'd produced five lives from it. I suppose I looked at it as another body sort of thing. But I sort of had my grieving for it and I ... I'd get bouts of depression and I'd start to grieve for the loss of this uterus.

While previous studies suggest Debbie's response is unusual, her view was shared by one of Ryan's (1985) respondents in her Melbourne-based study of 30 consecutive hysterectomy patient admissions. When interviewed pre-operatively this respondent commented that she did not want to lose 'the little nest where my babies grew' (p. 72). Amongst the women I interviewed, Debbie's account was unusual in another way, for she saw her womb as belonging not just to her but also to her husband, with the only difference being that she was 'the one taking care of it'. She recalled how her husband had grieved with her for loss of their joint reproductive years. This sharing of the emotional aspects of hysterectomy between Debbie and her husband was not reported by the other women I interviewed. In general, the men, according to their partners, did not appear to engage in any personal way with the loss of their partners' wombs. It was part of her body, and, if she experienced a sense of loss, her partner attempted to support her as best he could. However, as already noted, men's ability to provide emotional support was often limited.

Other respondents valued their wombs very differently, seeing them as things that could be discarded unceremoniously, with no ritual to mark their passing. Once its reproductive role was over, the womb had no material or symbolic importance:

> It [a womb] was just something that you had. It was just there. You had it; it functioned until it stopped functioning. (Rae, age 45)

> So, um having had my children I felt my uterus had done its job anyway, so I could get rid of it. I mean I always think if you had this little zip, once you've had all your children you could just chuck it away because it's basically a useless organ after that. (Lynette, age 38)

Lynette's comment bears a marked similarity to the ideas expressed by some gynaecologists, that the only function of the womb is reproduction, and when that is completed it 'becomes a useless, bleeding ... organ and therefore should be removed' (Wright 1969: 561). Her views were echoed by Vicky (44) who described herself as 'just minus a bit, but it became an incidental, unnecessary bit' after she had completed her family.

Loss of Femininity or 'Womanhood'

> I don't feel as though I've lost half my femaleness or womanliness or anything. [...] I am me, inside here I am me and I'm the me that I've always been. Yeah, no I don't see it as having any influence on me as a female, a woman I should say, whatsoever. (Vicky)

Like Vicky, many of the respondents rejected any suggestion that their hysterectomies had resulted in the loss of their femininity or impacted on their sense of identity. Janet echoed Vicky's words when she said: 'I'm the person I am'. The loss of her womb had made no difference to who she was, although she was aware that other women might have 'hang-ups' about the impact of hysterectomy on their self-image. That hysterectomy could have negative consequences for a woman's femininity was certainly part of the folklore surrounding the operation in some of the women's social networks. Cathy, for example, recounted how people used to question her after her hysterectomy as to whether she still felt 'like a female and all this sort of thing', to which she would reply:

> 'I don't feel any different'. And the fact that I was breastfeeding, I had this baby, you know ... well, I mean, of course I was a woman – I was breastfeeding! I had this baby, so I mean I had nothing to prove.

Jenny (37) too felt that she 'had nothing to prove' because, while being the mother of two young children, this was only one part of her self-identity:

I had never been you know, sort of strongly worried about my reproductive ability, it was (just) something that went along with being a woman. And I think because I'd always worked or studied or both, that I was, had a different perspective of myself you know, as a mother that was only one part of my whole perspective.

While Debbie grieved for the loss of the womb that had given her five children, she did not see the loss as affecting her femininity in any way or making her feel less of a woman. Although she had been concerned about this before the operation, her concerns had proved groundless:

I look at myself and I look at [other] women and I think, well I'm no different than you, I'm exactly ... I still feel very much a woman, you know what I mean ... I'm still very feminine.

Debbie's perception of her womb as 'another body' (see above) would seem to imply some kind of separation from self, which may have safeguarded her from the impact of its loss on her core sense of herself as a woman. However, her husband's ability to give her the emotional support she needed may also have been important. As she herself noted, men's response to their partners' hysterectomies could have a crucial impact on their feelings about themselves afterwards, 'because some males could make you feel that you weren't a woman'. Talking about her own, very different, experience, she said:

you need to know that regardless of whether you've had your reproductive organs taken out you're still in their eyes very much the woman that they married ... and you really need that reassurance from them. My husband was very good, he gave me that.

There was only one woman I interviewed for whom the value of her womb extended to her sense of herself as a woman; this was Catherine. For her, the loss of the 'security of having one's self intact' (Sloan 1978: 603) was devastating:

[I felt] as though a vital part of me had been removed. And it's more than just an organ far more than that; it was part of my femininity I suppose. I celebrated being female; sometimes I hated it but I celebrated it overall, and it was, there's some connection with it.

Having moved beyond talk of organs and reproductive ability, Catherine found herself searching for words that would adequately express her feelings about the loss of her womb:

I'm trying to think of exactly what it means but it's far more than that [a reproductive organ], it's a symbol I suppose. It had some value for me, whether that was qualities of being feminine, of being soft, of being life producing, of fertility – that's the word. But it went deep, it really went very deep this ... I don't really, I can't find it, exactly what it meant.

Once again Catherine finds that the language she has at her disposal is inadequate and she lapses into silence. Cameron (1985), Code (1991) and Spender

(1980) have all noted the inadequacy of masculinist language for describing uniquely female experiences. However, what is clear is that Catherine's womb, and the fertility that it symbolised, were intimately bound up with her sense of herself as a woman. As Drellich and Bieber (1958, cited in Hollender 1969: 14) noted in an early study, 'for some women, even though they desired no more children ... the loss of childbearing ability was viewed as rendering a woman something less than a complete female'. Catherine was one such woman.

While Lynette said that she had 'no sort of sentimental attachment [to her womb] because the problems it was causing ... were outweighing those sort of psychological things', she recalled that her response had not always been so matter-of-fact. At 32 years of age, with a young family and 'going through a very difficult stage in [her] marriage', she had rejected the idea of having a hysterectomy precisely because she was concerned about how she would feel without a uterus and how it would impact on her sexuality and femininity. Six years later, divorced and with a new 'very caring' partner in her life who helped her 'resolve a lot of issues', she had no such concerns.

'Mourning the Loss' or 'No Regrets': Understanding the Women's Emotions

In exploring respondents' experiences of hysterectomy an analysis has been presented which highlights the importance of the social context of women's lives and indicates that the meaning of hysterectomy varies across the life course.[6] Amongst the diversity of experiences, however, it is possible to broadly identify two contrasting emotional responses to the loss of the womb. Amy, Catherine, Dael, Debbie, Jasmine and Sarah all described periods of grieving as they 'mourned the loss' of their ability to reproduce or, more rarely, the womb as a valued reproductive organ and/or their femininity and 'womanhood'. However, the other women I interviewed did not generally share this response. Unlike Catherine for whom the womb was a symbol of her fertility and an integral part of her femininity, and Debbie for whom it was a living thing to which she had become emotionally attached, these women appeared to take a very functional perspective. The main function of the womb was reproduction. Since they did not wish to have any more children, the loss of their wombs was something that caused them 'no regrets'.

While age appeared to be a key factor, with the younger women being those most likely to experience a period of grieving for the loss of childbearing, other biographical and social characteristics influenced the likelihood of 'mourning the loss'. A conscious wish for more children, having young children (with the possibility of having another child if anything should happen to one of them), having friends who were still having children, and the possibility of entering a new relationship (and wanting to start a new family) all increased the likelihood of grief. While the older women did not share the younger women's feelings of grief for the loss of their reproductive functions (either real or symbolic), their stories suggested that this was because their lives had 'moved on' to a position where they had developed 'alternative identities', and the loss of their wombs was not experienced as a biographical disruption. Several participants spoke of having delayed the decision to have surgery until they had reached the stage when they no longer even fantasised about having another baby. By this time the youngest child in the family was often a teenager and the women themselves were more likely to be in paid work. Having found alternative sources of identity outside of motherhood, the womb was now seen as having no functional or symbolic importance; it was merely a body part that was causing problems, which its removal would fix. For these women it seems that the internalisation of the functional or medical model of the body may actually safeguard them from experiencing negative emotional responses to hysterectomy.

While limited by the relative homogeneity of its participants in terms of their cultural background, marital and parental status, the findings of this research suggest that a life course approach may be a useful way of understanding the meanings which women attach to hysterectomy and their emotional responses to the surgery. Further cross-cultural research that includes never married and childless women is needed.

Notes

1   It is estimated that one in five Australian women will have had a hysterectomy by the age of 50 (Santow and Bracher 1992) compared with one in four New Zealand women (Coney and Potter 1990), and one in five women in the UK by the age of 55 (Vessey *et al.* 1992). More recent figures for Australia, using the Australian Longitudinal Study on Women's Health, estimate a 22 per cent prevalence rate for women aged 45-50 (Byles, Mishra and Schofield, 2000).

2   The names used in this chapter are pseudonyms chosen by the women themselves. Amongst the more traditional girls' names is one boy's name (Dael) that had special significance for the respondent, and one name (Demon) that identifies the respondent as a fan of her favourite (Australian Rules) football team.

3   In such situations a hysterectomy is usually only one of a range of treatment options (which may include hormone or other drug treatments and, more recently, less invasive surgery such as endometrial resection or ablation). Surgery for 'benign' conditions is therefore usually referred to in the medical literature as 'elective' or 'discretionary', rather than 'indicated' (as in the case of cancer). However, the assumption that all the women in this study made a personal decision to have a hysterectomy is not quite accurate. For four of the participants the decision was effectively made by their gynaecologists as the following summaries explain. One woman chose to have an endometrial ablation, but there were problems during the operation and the surgeon made the decision to do a hysterectomy without her permission (Catherine). A second woman was diagnosed as having ovarian cancer and her surgery was therefore 'indicated' rather then 'elective' (Elizabeth); the diagnosis turned out to wrong. One woman who had suffered a prolapse (Kerrie) and another who had a fibroid polyp (Anne) were given no alternative treatment options.

4   The first time a woman's name is mentioned the age at which she had her hysterctomy is given in brackets. The majority of respondents (17) had abdominal surgery, with just three having a vaginal hysterectomy. Only one woman retained her cervix after the operation, while all but four retained their ovaries.

5 The category of heavy (or dysfunctional) bleeding and/or pain is a catch-all category where no specific medical reason for the symptoms can be identified.

6 It is important to note in this context that when I asked the women for their 'overall feelings' about having had a hysterectomy, even most of the women who had initially 'mourned the loss' said that they did not regret it at all (Jasmine) or that it was 'the best thing' they had ever done/had ever happened to them (Debbie, Amy, Dael). Only Sarah and Catherine still viewed the surgery as a 'tragedy' and a continuing source of grief. There were also three women who I would categorise as 'neutral' in terms of their overall response. While they had not experienced any strong emotional responses to the loss and 'accepted' that they needed to have it done, they would still have preferred not to have had surgery (Anne, Elizabeth, Rae).

References

Abel, E.K. (1990) Family care of the frail elderly. In Abel, E.K. and Nelson, M.K. (eds) *Circles of Care: Work and identity in Women's Lives.* Albany: State University of New York Press.

Abbott, P. and Wallace, C. (1997) *An Introduction to Sociology: Feminist Perspectives.* Second Edition. London: Routledge.

Alexander, D.A., Atherton Naji, A., Pinion, S.B., Mollison, J., Kitchener, H.C., Parkin, D.E., Abramovich, D.R. and Russell, I.T. (1996) Randomised trial comparing hysterectomy with endometrial ablation for dysfunctional uterine bleeding: psychiatric and psychological aspects, *British Medical Journal*, 312, 280-284.

Barglow, P., Gunther, M.S. and Johnson, A. (1965) Hysterectomy and tubal ligation: a psychiatric comparison, *Obstetrics and Gynecology*, 25, 520-525.

Bell, S. (1988) Becoming a political woman: the reconstruction and interpretation of experience through stories. In Todd, A.D. and Fisher, S. (eds) *Gender and Discourse: The Power of Talk.* Norwood, N.J.: Ablex Publishing Corporation.

Bernhard, L. (1992) Men's views about hysterectomies and women who have them, *IMAGE: Journal of Nursing Scholarship*, 24, 3, 177-181.

Birke, L. (1999) *Feminism and the Biological Body.* Edinburgh: Edinburgh University Press.

Brown, S., Lumley, J., Small, R. and Astbury, J. (1994) *Missing Voices: The Experience of Motherhood.* Melbourne: Oxford University Press.

Bury, M. (1982) Chronic illness as biographical disruption, *Sociology of Health and Illness*, 4, 2, 167-182.

Byles, J.E., Mishra, G. and Schofield, M. (2000) Factors associated with hysterectomy among women in Australia. *Health and Place*, 6, 301-308.

Cameron, D. (1985) *Feminism and Linguistic Theory.* London: Macmillan.

Carlson, K.J., Nichols, D.H. and Schiff, I. (1993) Indications for hysterectomy, *The New England Journal of Medicine*, 328, 12, 856-860.

Code, L. (1991) *What Can She Know?: Feminist Theory and the Construction of Knowledge.* Ithaca: Cornell University Press.

Cohen, S.M., Hollingsworth, A.O. and Rubin, M. (1989) Another look at psychologic complications of hysterectomy, *IMAGE: Journal of Nursing Scholarship*, 21, 1, 51-53.

Coney, S. and Potter, L. (1990) *Hysterectomy*, Auckland: Heinemann Reed.

Davies, J.E. and Doyle, P.M. (2002) Quality of life studies in unselected gynaecological outpatients and inpatients before and after hysterectomy. *Journal of Obstetrics and Gynaecology*, 22, 5, 523-526.

Dennerstein, L., Shelley, J., A.M.A. Smith and Ryan, M. (1994) Hysterectomy experience among mid-aged Australian women, *The Medical Journal of Australia*, 161, 311-13.

Drummond, J. and Field, P. (1984) Emotional and sexual sequelae following hysterectomy, *Health Care for Women International*, 5, 261-71.

Farquhar, C.M., Sadler, L., Harvey, S., McDougall, J., Yazdi, G. and Meuli, K. (2002) A prospective study of the short-term outcomes of hysterectomy with and without oophorectomy, *Australian and New Zealand Journal of Obstetrics and Gynaecology*, 42, 2, 197-204.

Fleming, V. (2003) Hysterectomy: a case study of one woman's experience, *Journal of Advanced Nursing*, 44, 6, 575-582.

Garry, R., Fountain, J., Mason, S., Napp, V., Brown, J., Hawe, J., Clayton, R., Abbott, J., Phillips, G., Whittaker, M., Lilford, R. and Bridgman, S. (2004) The eVALuate study: two parallel randomised trials, one comparing laparoscopic with abdominal hysterectomy, the other comparing laparoscopic with vaginal hysterectomy, *British Medical Journal*, 328, 7432, 129-133.

Hallowell, N. and Lawton, J. (2002) Negotiating present and future selves: managing the risk of hereditary ovarian cancer by prophylactic surgery, *Health: An Interdisciplinary Journal for the Social Study of Health, Illness and Medicine*, 6, 4, 423-443.

Hollender, M.H. (1969) Hysterectomy and feelings of femininity, *Medical Aspects of Human Sexuality*, 3, 6-15.

Kinnick, V. and Leners, D. (1995) The hysterectomy experience, *Journal of Holistic Nursing*, 13, 2, 142-154.

Linenberger, H. and Cohen, S. (2004) From hysterectomy to historicity, *Health Care for Women International*, 25, 349-357.

Lupton, D. (1998) *The Emotional Self: A Sociocultural Exploration*. London: Sage.

Lupton, D. (2003) *Medicine as Culture: Illness, Disease and the Body in Western Societies*. Second Edition. London: Sage.

McKenzie, C. and Grant, K.A. (2000) Hysterectomy – the patient's view: a survey of outcomes of hysterectomy in a district hospital, *Journal of Obstetrics and Gynaecology*, 20, 4, 421-425.

Moscucci, O. (1990) *The Science of Woman: Gynaecology and Gender in England 1800-1929*. Cambridge: Cambridge University Press.

Nathorst-Boos, J. Fuchs, T. and von Schoultz, B. (1992) Consumer's attitude to hysterectomy, *Acta Obstetricia et Gynecologia Scandinavia*, 71, 230-234.

Rannestad, T., Eikeland, O., Helland, H. and Qvarnstrom, U. (2001) The quality of life in women suffering from gynaecological disorders is improved by means of hysterectomy, *Acta Obstetricia et Gynecologia Scandinavia*, 80, 46-51.

Raphael, B. (1972) The crisis of hysterectomy, *Australian and New Zealand Journal of Psychiatry*, 6, 106-115.

Reid, B.A., Aisbett, C.W., Jones, L.M., Mira, M., Muhlen-Schulte, L., Palmer, G., Reti, L. and Roberts, R. (2000) *Relative Utilisation Rates of Hysterectomy and Links to Diagnosis*. Canberra: Department of Health and Aged Care, Commonwealth of Australia.

Roberts, H. (1992) Introduction. In Roberts, H. (ed) *Women's Health Matters*. London: Routledge.

Roeske, N. (1978) Hysterectomy and other gynecological surgeries: a psychological view. In Notman, M.T. and Nadelson, C.C. (eds) *The Woman Patient: Medical and Psychological Interfaces, vol 1, Sexual and Reproductive Aspects of Women's Health Care*. New York: Plenum Press.

Rowland, R. (1988) *Woman Herself: A Transdisciplinary Perspective on Women's Identity*. Melbourne: Oxford University Press.

Russell, C. and Schofield, T. (1986) *Where It Hurts: An Introduction to Sociology for Health Workers*. Sydney: Allen and Unwin.

Ryan, M. (1985) *Psychosexual Aspects of Hysterectomy*. PhD thesis. University of Melbourne.

Ryan, M. (1997) Hysterectomy: social and psychological aspects, *Bailliere's Clinical Obstetrics and Gynaecology*, 11, 1, 23-36.

Santow, G. and Bracher, M. (1992) Correlates of hysterectomy in Australia, *Social Science and Medicine*, 34, 8, 929-942.

Schofield, M., Bennett, A., Redman, S., Walters, W.A.W. and Sanson-Fisher, R.W. (1991) Self-reported long-term outcomes of hysterectomy, *British Journal of Obstetrics and Gynaecology*, 98, 1129-1136.

Sculpher, M.J., Dwyer, N., Byford, S. and Stirrat, G.M. (1996) Randomised trial comparing hysterectomy and transcervical endometrial resection: effect on health related quality of life and costs two years after surgery, *British Journal of Obstetrics and Gynaecology*, 103, 142-149.

Singer, D. and Hunter, M. (1999) The experience of premature menopause: a thematic discourse analysis, *Journal of Reproductive and Infant Psychology*, 17, 1, 63-81.

Sloan, D. (1978) The emotional and psychological aspects of hysterectomy, *American Journal of Obstetrics and Gynecology*, 131, 6, 598-605.

Spender, D. (1980) *Man Made Language*. London: Routledge and Kegan Paul.

Sydie, R. (1987) *Natural Women, Cultured Men: A Feminist Perspective on Sociological Theory*. Milton Keynes: Open University Press.

Vessey, M., Villard-Mackintosh, L., McPherson, K., Coulter, A. and Yeates, D. (1992) The epidemiology of hysterectomy: findings in a large cohort study, *British Journal of Obstetrics and Gynaecology*, 99, 402-407.

Williams, S.J. (2000) Chronic illness as biographical disruption or biographical disruption as chronic illness? Reflections on a core concept, *Sociology of Health and Illness*, 22, 1, 40-67.

Wolff, J. (1990) *Feminine Sentences: Essays on Women and Culture*. Berkeley: University of California Press.

Wright, R.C. (1969) Hysterectomy: past, present and future, *Obstetrics and Gynecology*, 33, 4, 560-563.

Chapter 4

'Healing and Feeling': The Place of Emotions in Later Life

Christine Milligan, Amanda Bingley and Anthony Gatrell

Introduction

While there is a considerable literature about ageing and older people in contemporary western society (see for example Jeffreys, 1991; Bytheway, 1995; Hepworth, 1998), as Bytheway (1995) points out, such literatures tend to focus on either the physiological or the social and structural impacts of ageing and ageism. In the process, they either sideline and ignore the role of emotions in the lives of older people, or ascribe stereotypical roles to them as 'emotionally worthy' or 'emotionally disreputable' (Hepworth, 1998). So, while it may be 'emotionally worthy' for an older person to express stoicism in the face of pain and suffering, it is viewed as 'emotionally disreputable' for an older person to express their emotional needs in terms of their own sexuality. Yet emotions are an integral part of our daily lives and being older does not mean that the range of emotions an individual may experience is in any way diminished. Moreover, we would argue that place plays an important, but under-researched, part in facilitating or constraining the emotional lives of older people. In this chapter we illustrate this by drawing on recent research conducted in Carlisle, a city in the north west of England, to consider the place of emotions in later life (see Milligan *et al.*, 2004, for further details of the study). In particular, we are concerned with the ways in which public, familial and social spaces act to facilitate or inhibit the expression of emotion in later life. This, in turn, may affect an older person's health and well-being.

The Paradox of Well-being: Placing Emotions in Later Life

There is increasing recognition of the complexity of factors that contribute to variations in the health and well-being of older people. However, emotions in later life are often neglected, as illustrated by the following comments from informal carers about care practices in institutional settings.

> he'd become incontinent ... and oooh, how humiliating, he's sitting with this green gown on and the catheter hanging down out on the floor – he had no RUG over him and no pyjama coat – it was just – ooh, I just felt so HUMILIATED for him when I went in.

They were dreadful. One of them [a doctor] said to Graham, 'now if you don't get out of bed today I'm going to take a big stick to you'! ... its just the STRANGE way they treated him, like he wasn't a human being or somebody with a bit of intelligence, he was just 'that thing in the bed'. (Milligan, 2004)

The above accounts of elder care point not only to the sidelining of older people's emotional distress, but illustrate how they are, at times, treated as objects that generate tasks for those responsible for providing and delivering care. Objectification of the older person renders unnecessary any concern for their dignity and self-worth, undermining the status of both the care professional and the elderly patient as thinking, feeling and emotional individuals.

Such examples of elder care illustrate how emotions are composed of a set of interrelated reactions, responses and subjective experiences that arise as a consequence of social processes (Averill, 1986). Hepworth (1998) argues that emotions are essentially learned ways of responding to social situations, and as such, may be open to transformation over the life course in that they change with age and experience. While this in no way diminishes the range of emotions an older person may experience, the attribution of distinctive emotions to age and stages in the life course is a socially constructed method of maintaining boundaries between different groups of individuals. This helps to explain tensions between personal experiences and public perceptions of emotional roles, (Davis, 1979), and highlights two dimensions to human emotions, namely subjective experiences, typically understood as originating 'inside', and the socially prescribed expectations that come from 'outside' (Hepworth, 1998). The interplay between subjective experience and the outer world influence cultural understandings of what it is to be 'old' and emotions associated with ageing and older age.

Hepworth (1998: 179) describes as 'the mask of ageing' the separation of an older person's inner emotional subjective experience from their outer bodily emotional expression, and this separation provides space for the elaboration of the distinction between emotions deemed 'morally worthy' in older age and those deemed 'disreputable'. The end result is the public expression of an 'idealised' older age where social respect for older people becomes wrapped up in cultural images of 'respectable elderly emotions' (typically devoid of sexuality and characterised by stoicism). The expression 'I don't feel old' – commonly voiced amongst older people – indicates that an individual's emotional experience in later life does not necessarily correspond with the public expression of what constitutes an idealised older age. The social performance of 'respectable emotions' can, thus, result in older people self-consciously distancing their subjective feelings from their outward expression or performance of emotion.

The lived experience of older age and emotions is as a fluid process rather than as a static collection of clearly compartmentalised roles. Indeed, Jerrome (1992: 160) maintains that the absence of clearly defined roles in older age can result in both a sense of freedom and a sense of confusion: '[t]o be happy and make the best of things in spite of pain or hardship is a moral and social obligation attached to the status of old ... persons in our society'. Such complexities raise important questions about how older people manage and experience these tensions and the impacts on their health and well-being. In this chapter, we explore this issue by considering how places can be actively constructed in ways that facilitate an outward expression of older people's inner emotional selves.

Within geography, work on emotions is relatively new, and to date, very little indeed focuses specifically on the place of emotions in the lives of older people (but see chapters by Hepworth and Morris and Thomas in this volume). Perhaps of most significance to this chapter, is the work of Graham Rowles (1978). While the place of emotions in later life was not the primary focus of his work, he was, nevertheless, aware of the importance of emotional attachments to place amongst older people, noting that 'locations "live" by virtue of emotions they invoke within the individual', and that '[f]eelings about place may reflect sentiments ranging from dread to elation. Often they are amorphous, multifarious, or inchoate. On an intuitive level it is clear that feelings associated with place are an integral component of the participant's geographical experience' (Rowles, 1978: 174).

For Rowles (1978), emotional attachment to places could be characterised as:

a   immediate – highly situation specific and relevant for only a short duration;
b   temporary – of rather longer duration and often repetitive in character; or
c   permanent – where there is stability in a deeply ingrained emotional identification attached to place.

In addition, he noted that emotional attachment to places could be classified as personal (stemming directly from an individual's unique experience), or shared (involving the mediation of other persons in sustaining an intersubjectively experienced sense of place).

The place of emotions in older people's lives varies across time and space. For example, social spaces felt to be supportive and friendly by day may become dangerous and frightening by night (see Hubbard, this volume), and emotional attachments to place may vary according to with whom, if anyone, experiences are shared. Places can foster seemingly contradictory emotions. It is these variations in the geography of emotions among older people that we explore in the remainder of this chapter.

Cultivating Health and Emotional Well-being in Later Life

The study on which we draw investigated the potential benefits of communal gardening and social activity for maintaining the health and well-being of older people. While our study focused on those over 65 years of age (participants ranged in age from 65 to 91), it is important to recognise that older age cannot be defined, simply, in terms of chronological age but also involves both physiological and social age (Ginn and Arber, 1995). Further, ageing is not reducible to processes of physical decline as if these occurred in a bodily vacuum sealed off from the social contexts in which people live. Featherstone and Hepworth (1993), for example, point to stereotypes of older people, predominant within a western culture obsessed by youth, that act to stigmatise experiences of growing older. Young people, they argue, construct negative emotional stereotypes around older people, and use pejorative terms to describe them, such as 'wrinkly', 'crumbly', 'fogey', and 'geriatric' (Featherstone and Hepworth, 1993, p. 308). Talking to an older person is seen as a time-consuming duty rather than an activity that contributes to a relational development. Yet older people are not a homogenous group. Rather, like other age cohorts in society, their socio-economic circumstances, cultural background, gender, health and physiological functioning vary (e.g. Bond, Coleman and Peace, 1993; Bytheway, 1995). This is reflected in our own study, which included participants from a variety of socio-economic backgrounds and ranged from the highly active (e.g. a 69-year-old male weight lifter, and a sprightly 89-year-old woman who regularly walks and participates in a variety of activities from the church to a cookery club) to those with severe mental and physical impairments (e.g. a 72-year-old female schizophrenic, and a 73-year-old man able to walk only very short distances without support).

Feeling Out of Place in Public Spaces?

In the late 1970s, Rowles commented that in a youth-oriented society, older people are an 'ever present reminder of mortality' and hence are 'ignored, shunned, or at best treated with indifference' (1978 p. 26). Nearly twenty years later, Bytheway (1995) maintained that one of the most invidious external constraints upon older people is the pervasiveness of negative societal attitudes. It would appear, then, that little has changed in the marginalisation of older people in western society. The persistence of ageism was evident in our own study. Thus, although some of the older people we interviewed spoke of the pleasure they derived from the often unexpected courtesies they received from younger people, and others maintained that the attitudes they encountered in public places were little different to those experienced by most other adults in society, experiences of ageism were also very common. For example, Sarah noted: 'When they know your age – when I broke my arm I had to go for therapy afterwards, and I went in [to the hospital] and they said "Are you Sarah Smith?" and I said, "Yeah". "Oh, I expected someone coming on a stick". You know? Because I was 86 at the time they were amazed because I was coming in under my own steam!' Others highlighted a tendency to associate physical decline amongst older people with decline in their mental abilities. In speaking of his profound deafness in one ear, Cyril remarked, 'I feel that people don't treat me as if I'm "all there" – sometimes if I can't hear what they are saying – and my brain functions alright, but people don't realise that!'

The marginalisation of older people is not manifest solely through their social interactions with other [younger] members of society, but is also a feature of the built environment itself. The mental and physical limitations often attached to the ageing process mean that some older people are subject to exclusionary processes very similar to those experienced by disabled people (see for example, Gleeson, 1996; and Imrie, 1996). Ralph, for example, has arthritis and severe ulceration on his legs, and spoke of the difficulties he experienced in negotiating public places, 'Coming up here [the library] it would have been hard work for me, coming up them stairs, but I come up the escalator. Going into [shop], it's all stairs going up – they're hard work. If somebody comes – a young 'un comes up behind me, I'll say "Come on, you're quicker than me". But I don't expect people to help us. There's always someone worse off than yourself …'. Ralph's experience draws attention not only to the failure of some public spaces to accommodate the needs of older people with mobility impairments, but is also illustrative of a stoic acceptance of such impairments and therefore of an emotional response characterised as 'morally worthy' by Hepworth (1998).

The most negative experiences of public space were attached to those places dominated by children and young people. This was often experienced as a 'temporary' but repetitive, emotional experience in which older people avoid certain places at certain times. Many older people (both male and female) consciously avoided travelling on local buses during 'the school run'. Where schoolchildren were not immediate relatives, they were almost universally viewed as key contributors to the negative experiences of travel on public transport. As Alma commented, 'When the schools are out, it's terrible … there isn't a proper school bus, so they take the ordinary bus that everybody else takes, and its pandemonium, so I don't take that bus, I avoid that. I'll stay in a store for half an hour 'til I know they've all cleared away'.

For other respondents, the negative experience of engagement with young people in public places was contrasted with 'the good old days', when older people were (imagined to have been) treated with respect and consideration. By contrast, accounts of subjective experience of place in contemporary society emphasise experiences of needs and desires being sidelined or ignored, with consequent impacts on older people's emotional well-being (compare Goodlove, 1982).

This negative experience of public places was not only associated with travel on public transport at specific times of the day, but also with the experience and fear of crime in the more deprived neighbourhoods in which some of our respondents lived. Fear of crime amongst older people is well-documented (e.g. Smith, 1987; Painter, 1989; Pain, 1997, 2000) and as the following interview excerpt reveals, it constrains people's use of space in place- and time-specific ways.

> Millie: '… there's been a lot of drug problems and a lot of break-ins recently in the streets around here. I look across at an empty house now and there was a fire there at night with the break-ins. I seen two men looking about and my neighbour told me to keep my doors locked and bedroom window shut. Unnerves me a bit … I won't go out if it's unsafe. But I won't let it get to me.'

As Rowles (1978) put it, older people are liable to become 'prisoners of space'.

The negative experience of the deprived localities in which some of the older people in this study resided highlighted the importance of access to natural environments for people's emotional well-being. Among our respondents, some accessed and enjoyed the peace and tranquillity of the rural landscape (where they might, for example, participate in such activities as painting, photography, walking or simply gazing at the scenery [but compare Parr *et al.*, this volume]). Others accessed alternative environments closer to their own homes. As Ted commented,

'Most mornings I take the dog for a walk out. There's a clique of us, we meet up by the river ... I go for a walk, maybe three or four miles – it's a leisurely walk, you know? And we put the world to rights – it's a good stress reliever I would say.' Ted went on to emphasise the importance of non-urban landscapes to his sense of well-being, noting, 'Around the river is definitely better, yeah. Along the streets, that would be no fun at all – particularly with the traffic. I very rarely come up town.'

While some individuals remain active and mobile into advanced older age, older age is often linked to physiological decline, a reduction in income and social interaction, and withdrawal from public places (Rowles, 1978). This withdrawal is often accompanied by a progressive constriction of older people's geographical life-space, which may be offset or intensified by social and environmental factors. In our study, for example, long dark nights or poor weather reduced respondents' geographical life-space. Older women in particular tended to retreat from environments beyond their homes at specific times of the day or year. So while Ruth, too, enjoyed walking along the river, it was a pleasure available to her only during daylight hours.

As noted above, Rowles (1978) characterised older people's emotional attachment to place as immediate, temporary or permanent. The evidence emerging from our study suggests that negative experiences of public places might result in temporary emotional *detachment* from place. By this we mean that while older people may have deeply ingrained emotional attachments to places or neighbourhoods, some seek to change their everyday routines and activities by removing themselves from specific public places at specific times, often as a consequence of the ageist behaviours of others.

Emotion, Family and Home

Finch and Mason (1993) pointed to the potential importance of family in the provision of emotional and moral support within the home, a role to which the family would appear peculiarly well suited given the often personal and intimate nature of anxieties for which people seek emotional support. However, given changes in the age structure and dependency ratio in contemporary society, there is a tendency to characterise this familial support in terms of older people being primarily recipients of support provided by their younger kin. Yet while much research confirms the importance of family relationships for older people (see for example, Nolan *et al.*, 2001; Vollenwyder *et al.*, 2002), data from our study indicate that older people are often givers rather than recipients of support. As Finch (1989) argued, older people often express considerable resistance to the notion that the dependency relationship is an asymmetrical one, and are anxious not to become 'too dependent' on their children and other relatives. As our study illustrates, older people (particularly women) can offer significant levels of support to others, often at the expense of their own emotional well-being. For example, Phillip describes the extensive support his wife, who has had two major operations and two heart attacks, gives to their adult son, who is both physically and mentally capable of looking after himself:

> the wife does a full-time job looking after two houses. For her age and her health you know, it's far too much, but she won't pack it in ... She hasn't got time to do her work and the housework at the son's house you see ... She makes him a bit of a meal seven days a week, does all his washing, cleaning, his bed, sorting the house out, putting new curtains up, taking them down – whatever. She does the lot. So in actual fact, she's looking after two houses! And that reacts back on me, of course, because I would like to get out more, but I can't.

Lorna also spoke of the tensions that arose from having her unemployed adult son live with her, and highlighted the importance of having friendship networks outside the home where she could gain emotional release:

> Everything I do, he criticises. It really is awful. If I go to Tesco's as I do every Monday, he's emptying the bag trying to see everything I've bought and if I've bought too much! ... It's truly dreadful, I'm seriously glad that I have my friends that I can get out to. He complains about absolutely everything, and it's very hard. It does get me down.

As well as undertaking housekeeping activities for adult children, other participants assumed regular caring, childminding, and babysitting roles within the family. This illustrates Jerrome's (1990) argument that age is often used by younger people as a basis for allocating jobs and roles to their elders on the assumption that older people have plenty of spare time, need occupation and are not the best judges of their own capabilities. As a consequence, older people may perceive that their emotional responses are either disregarded or ignored within their familial networks. This is particularly true for older women who often find themselves channelled into care and support roles that leaves little time in which to address their own emotional needs. Natalie, for example, expressed considerable bitterness at family expectations that she would care for her ageing family and relatives since being widowed some 15 years ago: she looks after an ailing brother and sister, and three years earlier, helped to care for another ailing sister. Although she often felt like 'the unpaid housekeeper', she commented that if she did not take on these tasks she would feel guilty and self-reproachful.

Declining health, or loss of a partner through bereavement, is a common experience for older people, and can prompt a range of feelings from isolation and depression to anger and guilt. Where an older person is caring for a spouse, sibling, or adult child the domestic situation may make it difficult for them to articulate their emotions. As Claire explained, the caring role she undertook left her no space within which she could express her own emotional needs:

> I looked after my husband for a long time, and then my mother died and I'd to go straight away and look after this auntie, she was 99 and had lived on her own, and she'd no family ... I hadn't time for the grief process. ... Sometimes with my aunt, I feel resentful towards her, because she's so clingy.

Revelations about ways in which both the physical and the emotional needs of older people were ignored within some families were sometimes shocking. However, we also found examples of very good family relationships characterised by high levels of reciprocity. Ted, for example, spoke of the care and emotional support he and his wife gave and received from his adult children and grandchildren, and of the close relationships that exist within the wider family network:

> ... probably unique these days, is that on a Sunday we have anything from six to eight others in our house for a sit-down dinner – and that doesn't happen very often nowadays. And on a Wednesday, we're up at the niece's and we sit around their table – I think it's good, because most people are sitting eating on their knee, watching the television and I think they're losing something there.

For others, childminding activities were a pleasure enjoyed equally by the older person, parent and child. In other instances adult children encouraged the older person to move closer to their own homes in order to facilitate their ability to support their ageing parent/s.

Thus, familial relationships vary, with some older people enjoying close and mutually supportive relationships with younger relatives, while others can feel drained by, and resentful of, expectations embedded within families. These latter often lacked any outlet though which to express their own emotional needs.

Inclusion, Emotion and Social Spaces

Jerrome (1990) argues that, following retirement, the absence of paid employment or other externally recognised social contributions reduces the status of older people, who are seen as having little of cultural value to trade with the young in return for support. This is especially the case in individualistic societies that emphasise the importance of self-reliance. As Ted commented, 'If you're old, you're pushed to the side and that's it, you know? ... older people have a lot to give, I think a lot of places [countries] realise that, but here, oh, you're old – you're pushed to the side'. Older people are thus rendered 'socially invisible', their relationships with younger members of society assumed to be ones of dependency and a lack of reciprocity. While we have argued that reciprocity flourishes in some family networks, there is, nevertheless, a perceived reduction in the status of older people within wider society. In these circumstances, peer group relationships increase in importance, and members of an age cohort often become a source of moral, practical and emotional support.

The increased likelihood of ill-health and mortality amongst older people, however, can also result in a significant shrinking of the peer group networks available. As Sarah remarked, 'I don't have as many friends, as you get older they seem to fall by the wayside ... They've all died, that's me, I've none left'. Furthermore, the loss or long-term illness of a partner can result in older people having to re-think their identities within their existing social networks. Women, in particular, noted their exclusion from social events, such as dances, dining out, holidays etc. where participation with a partner is the norm. Hence, as the focus group excerpt below suggests, the provision of spaces in which older people can meet and share experiences while participating in mutually enjoyable activities, serve to facilitate the forging of new social identities within new social networks.

> Esther: Well, I find that's the best bit about it, it [the social club] does get you out ... instead of sitting at home feeling sorry for yourself or depressed – it gets you out and about. You meet other people and you realise that other people have illnesses and things.
> [group agreement, some laughter]
> Rachel: And see, it's a different atmosphere when you're with youngsters to your own age group.
> Esther: Yeah. I feel you relax more with your own age group, you know? The pressure isn't there.
> Sally: Well, like yesterday, I mean we discussed other people's problems, and well everybody has some idea about doing something.

Such activity is particularly beneficial for those older people who live alone and/or who feel socially isolated. Within our study, those who were most isolated socially had also suffered substantial emotional ill-health. Tilly, for example, had suffered several bouts of depression over a number of years linked to social isolation arising from her husband's ill-health. For her, the social club represented a space in which she could meet and make friends with people of her own age. Discussing his experience of the gardening club, Stuart also maintained that, 'apart from the enjoyment of the allotment and the trips out, we have gained new friends. There have been a number of situations where members have helped each other in activities outside the club work.' Thus, social spaces in which older people can meet together, including social clubs and gardening clubs, offer opportunities for social contact and the development of new social networks. Supportiveness and reciprocity within these social networks may, we suggest, act to cushion the effects of stress, anxiety and negative emotional experiences often associated with ageing. Such networks provide a structure for acquiring new skills and knowledge, and a mechanism for enhancing a person's sense of self (Nolan, 1995; Langford et al., 1997).

There was a clear sense among our respondents that the social club served as an 'emotionally textured space', where for a short time each week, participants had a place in which they could express, share and have validated their emotions (in both explicit and sometimes implicit ways) within their peer group. Activities were supportive of this emotional process, offering a social context in which they could share common tasks and activities such as listening to invited speakers, experiencing new activities such as arts and crafts, and visiting new places on organised outings. As fellow club members got to know each other, they began, increasingly, to look forward to opportunities to talk, share stories, memories, histories and feelings about events. As Jerrome (1990) argued, informal rituals, including the making of tea, applauding of speakers, sharing of memories and so forth, all act to underline values and express important allegiances to the age group, the community of the club, friends and members of their social network. Such recurrent rituals and modes of participation provide security and a sense of personal connection between the individual and the collective. Lorna, in particular, noted the sense of enjoyment and satisfaction she gained from participating in a number of clubs following the death of her husband, commenting, 'I enjoy joining in, you know?'.

While there was evidence that new social contacts made within the social club were beginning to spill over into those wider life-spaces occupied by our participants, it was also apparent that developing social networks capable of providing emotional support, requires trust and friendship to be built up over time. Rachel, for example, described how her attendance at a hand-chime club over a four-year period had now developed into wider social activity where fellow members came together for conversation, meals, lunches and birthday celebrations (compare Wenger 1984).

Although the social club provided an important space in which older people could express and acknowledge the emotional lives of their peers, it did not appear to facilitate such intense emotional engagement as that evident in the communal activity associated with the gardening club. Participants gained social benefits from communal gardening, but their involvement had additional emotional benefits. Observation of participants during communal gardening activity suggested that they often became deeply focused and absorbed in what they were doing (see Figure 4.1). At other times, this absorption was interspersed with lively exchanges with other members of the group, whether shared laughter, disagreements, discussion, or the sharing of ideas and knowledge (see Figure 4.2). Emotional involvement in such communal activity can thus act to heighten feelings of solidarity, and give older people a sensory experience of continuing existence and vitality (Jerrome, 1990).

Participants expressed profound satisfaction in keeping the garden tidy, in nurturing seeds, in harvesting a good crop or successfully growing flowers, and in eating or giving the fruits of their labours to friends and relatives. As Alma commented, 'I came away with potatoes and lettuce, onions and mint – all our own home-grown produce … I know I'm very limited in what I can do, but I think getting the results gives you a boost … I take pride in the results'. These emotional responses were often deeply felt and powerfully expressed. Reflecting on her experience of communal gardening, Avril, in particular, noted, 'I think it [gardening activity] is therapeutic … when I've been round and seen all the things that are growing and talked to other people, I feel better when I come back [home].' Avril goes on to add, 'I think if I hadn't had that to look forward to I'd have been much more depressed and weary than I've been … It's been my main activity since I've had to give things up through being limited'.

While emotions are often individually experienced, they can also be shared, and may be consensually generated and reinforced (Rowles, 1978). It is, thus, possible to distinguish between an individual's unique experience of the physical and/or social context and shared feelings that are mediated by others in sustaining an intersubjectively experienced sense of place and identity. While these emotions can be permanent, comprising an intimate bond established over years of living in a particular place or neighbourhood, they can also be immediate or temporary –

that is, they may be time- and situation-specific, and may last for only a short duration. Within this framework, we suggest that the opening up of temporary, but supportive, spaces within which older people can express their emotions, particularly those who feel isolated by age differences within their family or social networks, can play an important role in sustaining their emotional well-being. As Susan noted, when recounting her feelings of loneliness arising from bereavement, 'I mean, it's a strange thing to say, but it can make you feel better talking to other people and you realise they are feeling the same way you are'.

Figure 4.1 Focused absorption in the gardening activity

Source: Photograph owned by authors

Figure 4.2 Social moments in the communal gardening activity

Source: Photograph owned by authors

The social and gardening clubs create spaces in which older people relax, share and express themselves in emotionally releasing and nurturing ways. The two clubs offer different kinds of space, requiring different kinds of investment from participants and generating different forms of emotional attachment. The space of the social club, for example, requires only the provision of a comfortable and accessible space in which older people can meet regularly and participate in their chosen activities. Conversely, the need to plan, work and nurture the communal garden draws older people into relationships with each other and with a particular place, more akin to those associated with the 'home', but without the complexities associated with familial relationships.

Concluding Comments

In this chapter, we have drawn attention to how ageism often results in a reduction in the emotional spaces available to older people. Ageist attitudes suggest that a full emotional life is the preserve of the young. Yet, as we have demonstrated, the emotional lives of older people are complex and multifaceted, influenced not only by declining health and the loss of partners and friends, but also by numerous other changes in familial and wider social networks.

What is also clear is that place plays an important role in facilitating or constraining the expression of emotion in later life. While we would not wish to homogenise their experiences, older people often feel marginalised in public places, their needs sidelined or ignored. For some, this results in reduced activity patterns and emotional detachment from familiar places and neighbourhoods. Ageism, however, is not confined to social interactions within public places, and while some older people enjoy close and mutually supportive family relationships, we also found evidence of an ageism within families, sometimes manifest in markedly asymmetrical patterns of physical and emotional support. Older women, in particular, are liable to find themselves providing substantial levels of support, often at the expense of their own emotional well-being, and with few outlets through which to express their own emotional needs. Thus, within public spaces and family relationships, ageist attitudes can press older people into 'emotionally worthy' social roles regardless of their underlying feelings. The kinds of social spaces discussed in the final section of this chapter can counter these patterns by offering inclusionary and supportive environments. These environments support the forging of new social identities within new social networks, providing spaces in which it is possible to drop the 'mask of ageing', and allowing older people to shift away from their socially prescribed roles to present and express themselves to their peers using the full range of their emotional experiences.

Acknowledgements

The study on which this chapter is based was funded by the UK Department of Health (former North and Yorkshire region) as part of the Healthy Ageing

Initiative. We would also like to acknowledge the contribution of the other members of the project team including Dr Rebecca Wagstaff (Public Health Consultant, and Director of Eden Valley PCT), Jessica Riddle (Age Concern) and Elizabeth Allnutt (Allotments Officer, Carlisle City Council). We would also like to thank the gardener and social club co-ordinator, Jane Barker, for her invaluable contribution to the project.

References

Acheson, D. (1998) Independent Enquiry into Inequalities in Health Report, HMSO, London.

Arber, S. and Ginn, J. (1995) *Connecting Gender and Ageing: A Sociological Approach*, Open University Press: Buckingham.

Averill, J.R. (1996) 'The acquisition of emotions during adulthood', in Harré, R. (ed.) *The Social Construction of Emotion*, Blackwell: Oxford.

Bond, J., Coleman, P. and Peace, S. (1993) *Ageing in Society: an introduction to social gerontology*, Sage, London. Second edition.

Bytheway, B. (1995) *Ageism*, Open University Press, Buckingham.

Davis, F. (1979) *Yearning for Yesterday,* Free Press: New York.

Featherstone, M. and Hepworth, M. (1993) 'Images of Ageing', in Bond, J., Coleman, P. and Peace, S. (eds) *Ageing in Society: an introduction to social gerontology*, Sage Publications, London. 2nd Edition. pp. 304-332.

Finch, J. (1989) *Family Obligations and Social Change*, Polity Press, Cambridge.

Finch, J. and Mason, J. (1993) *Negotiating Family Responsibilities*, Tavistock/Routledge, London.

Gleeson, B. (1996) 'A Geography for Disabled People?' Transactions of the Institute of British Geographers, 21, 387-396.

Hepworth, M. (1998) 'Ageing and Emotions', in Bendelow, G. and Williams, S.J. (eds) *Emotions in Social Life: Critical Themes and Contemporary Issues*, Routledge: London. pp. 173-189.

Imrie, R. (1996) Ableist geographers, disablist spaces: towards a reconstruction of Golledge's 'Geography and the Disabled', Transactions of the Institute of British Geographers, 21:2, 397-403.

Jerrome, D. (1990) 'Virtue and Vicissitude: The Role of Old People's Clubs', in Jeffreys, M. (ed.) *Growing Old in the Twentieth Century*, Routledge, London. pp. 151-165.

Jerrome, D. (1992) *Good Company: an anthropological study of old people in groups*, Edinburgh University Press: Edinburgh.

Milligan, C. (2004) Caring for Older People in New Zealand: Informal carers' experiences of the transition of care from the home to residential care, research report available at: http://www.lancs.ac.uk/users/ihr/staff/christinemilligan.htm.

Milligan, C., Gatrell, A. and Bingley, A. (2004) 'Cultivating health: therapeutic landscapes and older people in northern England', *Social Science and Medicine*, 58, 1781-1793.

Nolan, M., Davies, S. and Grant, G. (eds) (2001) *Working with older people and their families*, Open University Press, Philadelphia, PA.

O'Connor, P. and Brown, G. (1984) 'Supportive Relationships: fact or fancy?', *Journal of Social and Personal Relationships*, 1, 159-175.

Office of National Statistics (2001). *National Statistics Online – Census 2001* www.statistics.gov.uk/census2001/profiles.

Pain, R. (1997) 'Old age and ageism in urban research: the case of fear of crime', *International Journal of Urban and Regional Research* 21 (1) 117-128.

Pain, R. (2000) 'Place, social relations and the fear of crime: a review', *Progress in Human Geography*, 24:3, 365-387.

Painter, K. (1989) *Crime prevention and public lighting with a special focus on elderly people*, London, Centre for Criminology, Middlesex Polytechnic.

Rowles, G.D. (1978) *Prisoners of Space? Exploring the Geographical Experiences of Older People*, Westview Press, Boulder, Colorado.

Smith, S. 'Fear of crime: beyond a geography of deviance', *Progress in Human Geography*, 11, 1-23.

Vollenwyder, N., Bickel, J.F., dEpinay, C.L. and Maystre, C. (2002) 'The elderly and their families, 1979-94: changing networks and relationships', *Current Sociology*, 50:2, 263-280.

Wenger, C. (1984) *The Supportive Network*, Allen and Unwin, London.

Chapter 5

Guilty Pleasures of the Golden Arches: Mapping McDonald's in Narratives of Round-the-World Travel

Jennie Germann Molz

Introduction

Food and travelling are intertwined in various ways. Thomas Cook, the originator of modern tourism, declared that travel 'provides food for the mind' (cited in Brendon 1991, 31). It also provides food for the body. One of the ways travellers 'consume' the places they visit is by literally ingesting the foods available in those places (see Urry 1995 and in this volume on 'consuming places'). Food consumption recurs as a key theme in travel narratives, both in terms of tasting the 'other' and experiencing other cultures through food, and also in terms of eating familiar foods that signify home to the traveller. It is on this second aspect of food consumption that this chapter focuses. Drawing on narratives that travellers publish on-line while they travel around the world, this chapter considers the way travellers describe their patronage of globally franchised restaurants such as McDonald's.

The data extracts analysed in this chapter are drawn from a study of forty websites published by travellers while they were travelling around the world. While the travellers in the overall sample come from many different regions of the world, those travellers who describe eating at McDonald's tend to be from North America, the United Kingdom and Western Europe, and all of the extracts in this chapter are from North American travellers. Predominantly white, middle class, and from the first world, the travellers under consideration here clearly hail from a position of privilege in regard to processes of globalization, mobility and cultural consumption. McDonald's, with its familiar Golden Arches logo, is arguably a global icon, but, as I will discuss below, it is also an emblem of American culture and, for many, a symbol of Western cultural imperialism. In these particular travellers' stories, we can see the way processes such as globalization, global mobility, national identifications and the ability to feel at home are complexly intertwined and reproduced precisely through everyday activities such as eating at fast food restaurants.

McDonald's is ubiquitous, operating more than 30,000 franchises in 121 countries (McDonald's 2002). It is no surprise, then, that round-the-world travellers should encounter this restaurant and discussions of McDonald's crop up frequently on the websites travellers publish while journeying. This chapter explores emotional aspects of such encounters by investigating the paradoxical kinds of meanings and feelings that McDonald's evokes for those eating at (or avoiding) McDonald's. On the one hand, travellers acknowledge the global reach of McDonald's standardised food and environment. On the other hand, they detail the various localised menu offerings they order at McDonald's, such as beef-free Maharaja Burgers in Bombay, and in New Zealand the McKiwi Burger topped with beet slices. How does McDonald's alternate in travellers' stories between being a site of global homogeneity and a site of local specificity? Also, travellers talk about going to McDonald's to get a 'taste of home' even though, they claim, they never eat at McDonald's when they are at home. How, then, can something you never do at home come to mean 'home' when you are away? Indeed, how do these stories complicate the definition of home and its opposition to away? In other words, how do round-the-world travellers negotiate the apparent paradoxes of engaging with McDonald's as both a global space and as a local space and as both a homely space and as a foreign space?

McDonald's primary corporate model may be based on providing standardised products, services and environments, but travellers' emotional responses to McDonald's are far from standardised. As George Ritzer puts it in his account of the McDonaldization of society: 'There is a lot of emotional baggage wrapped up in McDonald's, which it has built on and exploited to create a large number of highly devoted customers. Their commitment to McDonald's is more emotional than it is rational, despite the fact that McDonald's built its position on rational principles' (Ritzer 1993, 149).

In fact, travellers don't always or only feel devotion to McDonald's. Their emotional responses to McDonald's are complicated and varied, ranging from desire and delight to disdain and disappointment – depending on how travellers define the space of McDonald's. In this chapter, I map out the mixed spatial meanings and mixed emotions surrounding McDonald's and similar global franchises. First, I describe the mix of emotions that travellers express about these places. Second, I show how these emotions map onto understandings of McDonald's space as alternately global and local, home and away. Finally, I interrogate the various identity and spatial strategies travellers employ to reconcile this mix of emotions and to justify their own patronage of McDonald's.

Mixed Emotions

Many of the travellers who publish websites on-line recognise McDonald's as some kind of forbidden pleasure and therefore their experiences at McDonald's elicit feelings of both shame and gratification. These travellers express embarrassment about eating at fast food restaurants. They say they are 'ashamed' and 'hate to admit' that they 'slipped' or were 'naughty' because they ate at McDonald's or Pizza Hut while away. One traveller on a small group tour in China describes in her website how she 'snuck' to McDonald's in Beijing when she was able to get away from the rest of the group. Of course, not all travellers struggle with ambivalent emotions about eating at fast food restaurants while travelling. Some travellers unabashedly embrace the opportunity to eat at McDonald's and they make no apologies for this. However, it is more common for travellers to talk about their meals at McDonald's, Pizza Hut and similar restaurants in terms of 'indulging', 'scarfing', 'gorging' and 'succumbing to temptation' – highlighting the shameful pleasure of overindulging in something they feel they're not supposed to have. The way travellers tend to use such terms, though, glosses over the acute feelings of shame and anxiety that, for many people, are tied into associations between overeating, eating disorders and obesity (see Coveney 2000; Canetti, Bachar and Berry 2002). It is important to point out that, unlike disclosures of guilt or shame that in other contexts (see for example Heenan's chapter in this volume) speak to intensely felt emotions that individuals may prefer to hide rather than make public, travellers' admissions of 'guilt' are, admittedly, very playful. Their appeal to a wider cultural ambivalence regarding both fast food restaurants and the desire to eat familiar food while travelling is framed more as amusing than truly shameful. Nevertheless, these travellers' stories about eating at McDonald's point up some of the complex emotional responses travellers have regarding the political and cultural implications of globalization and their own global mobility.

This is the case for Marie, a woman who travelled alone overland around the world in 2001. The following extract from a live chat she hosted on her website near the end of her trip demonstrates some of the paradoxical emotions tied up in McDonald's:

> J: Did you have a favorite and least favorite McDonald's experience? …
> Marie: J: My fave McDonald's moment was horribly politically incorrect. …
> W: Do tell!
> Marie: I … had been starving my way through Sudan due to Ramadan, not to mention vomiting for 2 days in Khartoum, and then I got to Cairo and McD's was empty due to Ramadan so it was just me and the Big Mac. Mmmmm … sounds awful to be eating McD's [sic] abroad (or at home, which I don't), but it was heavenly. … I know it isn't culturally tolerant to be scoffing down a big mac [sic] during Ramadan but I couldn't help myself.
> J: If it's okay for them to be open and cook it then it's okay for you to eat it.
> B: Vomiting for 2 days in Khartoum – sound like a winner title for a travel piece. Or Jihad vs McWorld. Stuffing down my Ramadan burger but waiting until dark by Marie (Marie's World Tour 2002).

During this chat, Marie concedes that eating at McDonald's is politically incorrect and culturally intolerant. With the allusion to Benjamin Barber's (1995) book Jihad v. McWorld, the on-line chatters acknowledge McDonald's as the cultural symbol of 'The West' as opposed to Islam and toy with the political tensions tied into eating at a McDonald's franchise in a Muslim country during the holy fasting month of Ramadan. Nevertheless, as Marie says, she could not resist the temptation to eat there. She also recognises that eating at McDonald's while travelling abroad 'sounds awful' – and here, she also claims that she never even eats McDonald's at home. But at the same time, given her circumstances, the Big Mac was 'heavenly'.

This recognition of McDonald's as a forbidden pleasure and the way travellers justify indulging in this pleasure are wrapped up in the ways McDonald's is symbolically spatialised in travellers' stories. These mixed emotions about McDonald's as something that is desired, but not, as one traveller puts it, 'the done thing', are related to the travellers' struggles to maintain a sense of an adventuring identity in a space where the boundaries between local and global, home and away sometimes fail to hold. To some extent, the sense of ambivalence that travellers express toward eating at McDonald's may reflect an implicit distinction travellers attempt to uphold between themselves as 'travellers' and other 'tourists'. Historically, figurative distinctions have been made between travellers and tourists to bolster the upper class/lower class distinctions that were threatened when cheap transportation and an accessible tourist infrastructure democratised travel in the nineteenth century (Urry 1990; Sharpley 1999). As Buzard (1993) explains, the 'real or the perceived encroachment of "tourism" on districts that had been the preserves of a privileged few nourished the urge to delineate' tourists from travellers (81). As a result, tourism became negatively associated with mass travel and packaged tours to predictable destinations while travel became associated with challenging journeys off the beaten track and authentic encounters with local cultures (see Boorstin 1961). Given such distinctions, 'inauthentic' activities such as eating at McDonald's, and indeed capitulating to the desire for familiar foods, may be seen as undermining the traveller's status as a 'traveller'. However, travellers' ambivalence about eating at McDonald's may also be related to the very fact of constant mobility that characterises their journeys. Unlike tourists who tend to travel to one or two places for a short period of time, round-the-world travellers are generally on the road for up to a year, incessantly moving from place to place as they circle the globe. In this case, travellers must find quick ways of feeling at home or forging emotional connections to places during their brief visits. They do so in several ways, one of which is resorting to McDonald's as a familiar symbol of homeliness. However, McDonald's does not just symbolise homeliness; it may, at the same time, symbolise the alienation wrought by global homogenisation of local spaces (see for example Augé's (1995) work on the non-places of supermodernity). How then does McDonald's map out in travellers' stories as both a site of familiarity and comfort and as a symbol of globalization?

Mapping McDonald's

Travellers project a variety of spatial meanings – often conflicting ones that blur the distinctions between local and global and between home and away – onto the Golden Arches. These spatial qualities do not inhere in McDonald's restaurants themselves; in other words, McDonald's is not inherently a global icon or a symbol of the local or an enclave of 'homeliness'. Instead, the way these spatialisations and related emotions are performed in each traveller's narrative depends on where the traveller comes from, on how far (geographically and emotionally) from home the traveller feels, on how different or familiar the location seems to be, and on the traveller's trajectory (where she is, has been, or wants to go).

The emotional dimensions of globalization – how people feel about globalization as well as how they 'feel' global – are commonly examined in terms of attachments and affiliations to cultural or national homes (see Robbins 1999). Many critics acknowledge the complicated status of home in the modern era of globalization and argue that the question of home becomes ever more complex for people on the move and for people living in a world on the move, as home and other 'grounds' of belonging become deterritorialised (Gupta and Ferguson 1992; Bauman 1998). While some experience this shift as a sense of homelessness, what Featherstone (1995, 1) refers to as an 'inability to find the way home', for others it manifests as a proliferation of homes (Ahmed 2000). For example, a privileged, mobile, elite with certain physical, financial, cultural and technological resources, such as the travellers under consideration here, are able to 'feel' at home wherever they are. Thus emotional responses to globalization are frequently enmeshed in a matrix of feelings about home.

Within this matrix, home is evoked multiply and contradictorily, for example as a sense of pride or shame or disappointment regarding one's national home, as a feeling of safety, comfort and familiarity derived through the perceived stasis of home, or as a desire to leave home precisely because it is perceived as static, banal or boring. For many of the travellers in this study, questions of home (or feeling at home), mobility and globalization merge precisely in the space of McDonald's. Travellers negotiate emotions of desire, disappointment, excitement and guilt through these mappings of McDonald's thereby expressing ambivalent sentiments upon encountering the real or imagined effects of globalization on the places they visit. As one traveller profiled on the popular on-line travel newsletter Boots 'n' All exclaims, 'You only live once, and there's a whole world out there waiting to be explored. Live a little and see it before the West globalizes every corner with a McDonalds and Starbucks' (Boots 'n' All 2002).

The negative emotions travellers express toward eating at McDonald's are in part related to travellers' association of McDonald's with corporate globalization and Western cultural imperialism. The Golden Arches is one of the most powerful and widely recognised brands in the world and Ronald McDonald is almost as well known as Santa Claus and Mickey Mouse (Watson 1997; Klein 2001; Schlosser 2002). McDonald's symbolises globalization both in its already ubiquitous presence throughout the world and in its corporate model of expansion – which has influenced a host of similar franchising corporations and indeed, according to Ritzer's work on McDonaldization, the way social institutions more generally are being organised (Ritzer 1993 and 2002; Urry 2000; Schlosser 2002). Some scholars as well as travellers also consider it a 'vehicle for potential cultural globalization' (Beynon and Dunkerley 2000, 24; Featherstone 1995; cf Watson 1997). Seen from this perspective as a globally, albeit undeniably Western, homogenising force that displaces and replaces local culture, McDonald's characterises not only globalization, but everything that is wrong with globalization. Aware of the effects of globalization on the landscapes and cultures of the destinations they visit, many travellers react to McDonald's with emotions of disappointment, disdain and even rejection, lamenting the American-esque 'wasteland' of McDonald's and strip malls in places like Australia's Gold Coast, for example. Steve, a traveller from

Canada, expresses the disappointment he and his girlfriend experienced when arriving in Australia, only to find the same franchised storefronts they had left behind in Canada. The journal entry in his website reads:

> Brisbane, Australia
> Just Like California
> We were at Wet 'n Wild, a waterslide theme park (very Wet, not at all Wild) just south of Brisbane, and one of several theme parks within a few square kilometers. It wasn't just the theme parks that seemed familiar, though: it was everything.
> It was the KFC's and Pizza Huts and McDonald's. The Nevada Bob's Discount Golf and the Target department stores. The eight-lane superhighway. What we were doing felt more like vacationing than travelling. There were no surprises, no adventures, no weird cultural gotchas – in fact, the biggest challenge for [us] was dealing with this familiarity. Why were we spending so much time here, in comfort and mild boredom, when exotic lands of adventure lay just to the north? (Big Adventures 2002)

In Steve's case, familiarity breeds contempt. In his quest for 'cultural gotchas' and 'exotic lands of adventure', Steve is trying to escape familiarity. And yet the global reach of corporate franchises reproduces a familiar landscape even far from home. These emotions of disappointment and even boredom are usually expressed in response to an external view of McDonald's – as part of the sprawling landscape of corporate globalization. But once inside the McDonald's, a different set of emotions comes to the surface.

McDonald's is not only an icon of the global; for many travellers, it represents the local as well. If some travellers find McDonald's global ubiquity depressing, others find enjoyment in the 'localization' as opposed to 'globalization' of McDonald's space (see Watson 1997). Once inside the McDonald's, travellers begin to detail the localized elements of the space. They are delighted by signs of the local culture on the menu and by the fact that they can actually 'eat local' at McDonald's. McDonald's thus becomes a site of 'culinary tourism' (Watson 1997, 38; Long 2003). Indeed, travellers often use this 'local flavour' as an excuse for indulging in the forbidden pleasures of McDonald's; they can still be adventurous eaters – ordering things like red bean pie, for example – even in the very safe environment of McDonald's.

McDonald's 'localization' of its menus has become an integral part of its marketing strategy. Playing on the way this strategy might tie into the McDonaldization of tourism, Ritzer and Liska (1997, 100-101) lay out the following hypothetical itinerary, tongue-in-cheek:

> We could even envision a world tour of McDonald's restaurants. ... In addition to mandatory visits to McDonald's in Moscow and Beijing, who wouldn't want to visit Norway and eat McLaks (grilled salmon sandwich with dill sauce on whole-grain bread); the Netherlands and devour a groenteburger (a vegetable burger); Uruguay and feast on McHuevos (hamburgers with poached egg) and McQuesos (toasted cheese sandwiches); and Japan where one can find a Chicken Tatsuta sandwich, fried chicken spiced with soy sauce and ginger, with cabbage and mustard mayonnaise? This itinerary is presented with a sense of irony, but on second thoughts our guess is that some clever tour operator could earn much income from such a tour.

Their predictions are not too far off the mark. When Salli and George took their two young daughters around the world in 1997, they made it the family's mission to eat at McDonald's in every country they visited. They write on their website: 'We have decided to try McDonald's wherever we go, partly to see if they have local dishes on the menu, and partly because the girls get homesick and need a break' (WorldHop 2001).

True to their word, this family visited about a dozen McDonald's while circumnavigating the globe. While these travellers frequently ate at local eateries, they also used their McDonald's experiences to try local foods offered on the menu and even to learn some of the local language – such as the Welsh words for burger, cheese and french fries. Throughout their website, they provide a rundown of the local offerings on McDonald's menus in Versailles, Amsterdam, Hong Kong, Cairo and Wales. Salli describes at length their experiences in the Bombay McDonald's:

> We walked into a small cement courtyard with a big statue of Ronald McDonald at one end and bright yellow and red painted chains and stanchions to keep the crowds orderly ...
>
> Once in the restaurant, we were greeted with signs proudly announcing 'no beef used here'. We ordered a Maharaja Burger (a double decker similar to a Big Mac, but made with ground mutton), a Veg Burger (made with a chick pea patty) ... and Veg McNuggets ... [W]e thought we'd try the stuff we usually can't get ... George also checked out the hot pie – which was not apple, but pineapple. Although we were pleased to try something new, I note how happy we were to get the familiar standard McDonald's-issue fries, shakes and cokes after our month in India. ...
>
> Although the meals are expensive by Indian standards, it is Indians – mostly the young and affluent – who eat here. The place was clean, air-conditioned, had an outside patio for those who could stand the heat, and the bathrooms were some of the best we encountered in India ... All-in-all pretty pleasant, and, at that point in our travels, a welcome escape (WorldHop 2001).

This extract reveals the family's ambition to try something new (as self-proclaimed adventurous travellers, a point they reiterate throughout the website) but also their relief at getting both food and space that was familiar and clean. The Bombay McDonald's is both globally standardised (the staff and customers are orderly, the fries and Coke are standard-issue, the bathrooms are clean), and local (they don't serve beef, they serve pineapple instead of apple pie, and most of the patrons are Indian), but also a home-like 'escape' after a month in India eating primarily local dishes in local establishments.

McDonald's is constructed as a local/global space evoking attendant emotions of disappointment about the over-familiarity of global franchise chains on the one hand, and delight in 'discovering' the local cuisine (as translated by McDonald's) on the other. But for many travellers, it is also constructed as a soothingly homey space. As much as this family learned from the 'glocalized' menu (see Robertson 1995), they admit that they ate at McDonald's for more than just educational purposes. They ate there 'to maintain their sanity' and because they felt homesick.

The travellers in this study equate McDonald's to home both explicitly and implicitly. Different travellers explicitly refer to McDonald's as a 'home comfort', as a 'taste of home' and as something that can remedy homesickness or 'temper' the 'home cravings'. McDonald's is a 'comfort zone' where travellers can indulge in 'comfort food' to ease the disorientation or difficulty of constant travel through foreign cultures. Travellers equate McDonald's to home implicitly by describing it as a space of familiarity, predictability, cleanliness, orderliness and belonging. For example, travellers from the United States gravitate to McDonald's or other recognisable restaurants to celebrate American holidays like Thanksgiving or Fourth of July (American Independence Day) with familiar foods. McDonald's becomes a space in which travellers who are geographically far from home can feel at home. In the following extract, Todd describes how he and his travelling companions, feel 'at home' in a Beijing McDonald's:

> Since we had gorged ourselves with wonderful Chinese food since our arrival, Jeff suggested McDonald's. He was curious to see if he felt just [as] at home in a Chinese McDonald's as he does at the Golden Arches two miles from his house.
>
> The language differences aside, the Big Mac in China turned out to be just as greasy and inconsequential as the Big Mac in the States ... We all kept our distinctly American posture in this American establishment, as Doug yelled across the open space of the restaurant to alert our host where we had chosen to sit ... We all used the restroom at Mickey D's and each of us kept a swatch of their beautiful six-ply fuchsia colored toilet paper ...
>
> [We left] with full bellies and warm thoughts of American cuisine (Drive Around the World: Latitude 2001).

Even in this somewhat cynical description of a trip to McDonald's, the sense of comfort and contentment of slipping into a familiar American posture and leaving with 'full bellies and warm thoughts' is almost palpable. This reference to the bathroom at McDonald's is not incidental. In fact, McDonald's bathrooms are famous among traveller circles for their dependability – they are usually clean and actually have toilet paper (see Watson 1997). As one traveller puts it, a 'McDonald's [bathroom] is usually a safe alternative anywhere in the world' (World Wander to the World Wonders 2001). If McDonald's is a place to experiment with bits of the local cuisine, it is also a culinary and spatial oasis from the local when it all gets to be too much. Travellers are relieved to find a clean, air-conditioned haven where they know what to expect, where the food always tastes the same, the environment always feels the same, and there is always toilet paper in the bathroom. It all sounds so comforting. Why, then, do travellers express feelings of guilt and embarrassment about eating at McDonald's while they are abroad? And how do they come to terms with these mixed emotions of shame and comfort?

Emotion, Space and Identity

The predicament lies in the equation of McDonald's with home and home with familiarity. Both Sara Ahmed (2000) and Anne-Marie Fortier (2003) have interrogated the parameters of this equation of home as familiarity. In asking what it means to 'be-at-home', Ahmed questions a particular discourse of home as a place that is so familiar and so comfortable that it must be 'overcome' or left behind. In this narrative, Ahmed (2000, 87) argues

> Home is implicitly constructed as purified space of belonging in which the subject is too comfortable to question the limits or borders of her or his experience, indeed, where the subject is so at ease that she or he does not think. ... To be at home is the absence of desire, and the absence of an engagement with others through which desire engenders movement across boundaries.

To Fortier (2003, 116) connecting home with familiarity 'entails stasis – it is a site where things and subjects stand still, and it is there to be left behind or desired'. Both writers argue for a reconfiguration of home as a site where strangeness, desire, and queerness – attributes otherwise associated with 'away' – are already encountered. The problem for the travellers from the United States whom I have cited here is, I think, a corollary of this argument. They are not struggling with a familiarity that is static, but with a familiarity that won't stay put; that can't be 'left behind'. The question for them is not so much one of encountering difference and 'away' at home, but rather encountering home while they are ostensibly 'away'. What happens when familiarity implies, not stasis, but rather movement? What happens when home can't be left behind or desired because enclaves of 'homeliness' encircle the globe?

The confusion of spatial boundaries between home and away is linked, as well, to the blurring of global and local spaces. The fact that the spaces of home and away are intersecting with each other leads some writers to argue that places around the world are becoming more like each other through processes of homogenisation and globalization (Rojek 1998; Holmes 2001). The result is that when people travel, they find that 'other places' are 'increasingly internal to the place from which [they] set out to do this travelling. ... The confrontation of "difference" and otherness (be it in virtual travel or embodied travel) is emptied out when we arrive at a destination to find that the object-worlds we had just left have followed us' (Holmes 2001, 10). According to this argument, as clearly demarcated places of 'home' appear in places that are supposed to be foreign, exotic or at least different, the individual is overwhelmed by familiarity and robbed of the transformational experiences promised by encounters with difference and 'otherness'. Rojek (1998, 38) explains that:

> In traveling, we reveal the limits of our own personal and cultural worldviews, as well as encountering customs, habits and values which differ from our own. Tourism shows us new ways of organizing personality and life space, and exposes the socially constructed character of our beliefs and values. It also carries a critical potential in contrasting the gnomic character of our routinized existence with the simultaneous worlds which are in reach merely by taking a car ride or buying a plane ticket.

Rojek's concern is that when leisure spaces become more like domestic spaces, and vice versa, through processes of commodification and globalization, they cease to provide the experience of contrast and difference central to this personal transformation. The accessibility of McDonald's almost anywhere they go sabotages travellers' ability to play out this narrative; they can't leave a home that seems to follow them around the world, and therefore the ability to 'overcome' home – with all of the transgressive and transformative potential that implies – is undermined.

Travellers' excuses for eating at McDonald's can be understood as attempts to renegotiate this narrative. As I have already mentioned, some travellers justify their patronage by trying local dishes – this injects a measure of 'away' or exoticness into a very familiar place. Others defend their patronage of these chains by claiming that they are 'recharging' in a western oasis before venturing back out into the unknown. Or, travellers declare McDonald's as a 'reward' for having endured travel in remote and difficult areas. They use the fact of being immersed in and literally 'fed up' with local culture and food as justification for eating at McDonald's. These justifications attempt to place the travellers solidly within an away space in opposition to the homeliness of McDonald's familiarity. But this doesn't always work.

As Ahmed (2000, 89) says, 'the boundary between home and away is permeable. … Movement away is always affective: it affects how "homely" one might feel or fail to feel'. When the distinction between home and away does not hold travellers find themselves in a bit of an emotional quandary. They regulate themselves against a host of 'shoulds' – what good, adventurous, culturally tolerant travellers should do – and then berate themselves for succumbing to temptation. Here, the temptation is home; to feel homely, if only for a little while amidst a sea of what is to them exotic difference. But the space of McDonald's is not so clear cut and its association as global or local, home or away is contingent on the travellers' own trajectory.

How travellers feel about McDonald's depends on where they are, where they have been and where they want to go. Consider the following three journal entries from a website by Vija, who travelled around the world with her husband from 1999 to 2000. As Vija explains, seeing an outlet like McDonald's could have had several different meanings for her; the meanings she ascribed to McDonald's here were contingent on the fact that she had already been travelling for eight months and was nearing the end of her trip. In Bangkok, Vija describes her excitement at encountering Western restaurants:

> We are in the land of McDonald's, Burger King, Pizza Hut, Famous Amos Cookies, A&W, Baskin & Robbins, Mrs. Field's Cookies, and even Dunkin' Donuts. Had I flown in directly from the U.S. in search of the exotic Bangkok I had pictured in my mind prior to coming here, I probably would have been disappointed. But not this time. Not this cowgirl. Not this born-and-bred-in-the-United-States-fast-food-loving-westerner. I was ecstatic, and in my mind mapped out our days here in Bangkok to see if we could sample every one of these culinary delights (Ben and Vija's RTW Adventure 2000).

After leaving Bangkok and arriving in Singapore, Vija is still delighted to find familiar places to eat. Here, she describes in very similar terms her relief at being in an 'American' cultural context as she discovers a Starbucks coffee shop in Singapore:

> You know, eight months ago when we started the trip I would have been disappointed in Singapore. If we would have flown straight here from the States, I would have been disappointed in it's [sic] sameness. There is virtually no exoticness or culture because it seems like the whole city is brand new. But, eight months later, I love it. And I think it's because Singapore is a mini-America. Perhaps my love affair began when I crossed the road from our hotel and stopped dead in my tracks in front of the first Starbucks I had seen on the trip. I practically ran to the counter and ordered a latte (Ben and Vija's RTW Adventure 2000).

Vija explains that her emotional reaction to encountering McDonald's, Pizza Hut and Starbucks is due in large part to the trajectory of her trip. The further she feels from home, the happier she is to encounter these 'homely' restaurants. She explains that she and her husband had originally intended to do the trip in reverse, but then changed their minds in order to save the more 'western' regions of the trip for the end:

> We were going to begin the trip in Fiji, Tahiti, NZ and Australia, then up SE Asia, and so on. We switched it for a couple of different reasons, one of them being that I thought we would appreciate the westernness [sic] of Australia, etc. more at the end of the trip. I couldn't have known how right that decision would be. At least as far as I'm concerned. Somebody else may have a different experience, but for me this was perfect. I've been embracing western conveniences, including fast food and shopping malls, ever since we got a taste in Bangkok. ... Call me what you want because I know the real truth: I'm honest. I have grown a lot on this trip, and I appreciate the scale and vastness of cultures and nations in a way I could not have even known before we came. However, I also know I am a product of my culture and it would be hard to give it up (Ben and Vija's RTW Adventure 2000).

As these extracts demonstrate, emotions of disappointment and emotions of delight are spatialised through the itinerary of the trip. At the beginning of the trip, when Vija was seeking difference and exotic-ness, McDonald's would have been too close to home to satisfy these desires and would have evoked disappointment instead of happiness. By the end of the trip, after enduring months of exotic-ness and difference, she was delighted to find signs of home marking the last leg of her trip. Further, Vija notes the expectations tied to spaces of home versus spaces of away. She acknowledges that going to American outlets like McDonald's or Starbucks may be perceived as too 'homely' to provide opportunities for growth and learning about the world. Vija responds to this imagined criticism by claiming that she has grown and learned; but she also acknowledges the almost immutable internalisation of her own American cultural identity. The fact that almost all of the travellers I have cited in this paper are from the United States is not coincidental. As many writers argue, McDonald's symbolises America, American culture and

American imperialism (Schlosser 2002; Ritzer 1993) so it would follow that American travellers equate McDonald's to their home.

Travellers from the United States justify their visits to McDonald's as an instinctual surrender to their cultural or national identity as Americans – like Vija above. Thus a reiteration of national identity and the emotions associated with this affiliation also shape travellers' consumption of McDonald's abroad. Indeed, another emotional response to McDonald's is one of national pride and even civic duty. One couple, delighted to find a Taco Bell (a U.S.-based Mexican food franchise) in Central America, makes this plea to their readers 'I am not sure if Taco Bell is going to make it in Ecuador but if you happen to be in Quito stop by for us, we would hate to see such a wonderful American experience not make it' (World Wander to the World Wonders 2001).

Interestingly, though, there are several examples of travellers who are not from America referring to McDonald's as a 'comfort of home' and as a familiar, homely space. This is not to say that everyone finds McDonald's and similar global franchises a familiar and comfortable space. For some travellers, these fast food spaces can be quite unfamiliar and even hostile. Marie, the traveller introduced earlier, describes her discomfort in a KFC in Indonesia. 'I was the only foreigner for miles, apparently, and when I walked to the KFC for my late-night super-spicy dinner, everyone stared unabashedly. Some men whistled, and lots of people hissed. I couldn't wait to get out of Indonesia' (Marie's World Tour 2002).

In this example, the American franchise fails to provide a safe, homely respite for the traveller. And in his book Fast Food Nation, Eric Schlosser (2002) lists several incidents of anti-American demonstrations carried out against McDonald's outlets in countries such as China, Denmark, Columbia and Russia. On the other hand, some places adopt McDonald's as their own and Schlosser quotes a McDonald's worker in Plauen, Germany saying that the people there do not consider McDonald's 'foreign'. In fact, as Watson (1997) argues, McDonald's franchises in places like Hong Kong, Beijing, Seoul and Tokyo cease to symbolise America or even 'foreign-ness'. Instead, McDonald's becomes as much a symbol of home to local patrons in Asia as it is to Americans abroad. This chapter has focussed on the emotional ties that American travellers have to McDonald's when they encounter this familiar restaurant far from home. But as Schlosser's and Watson's studies show, considering the way local consumers, as well as the people who work in McDonald's, negotiate their emotional responses to McDonald's traces out even more nuanced contours of the emotional landscape of McDonald's.

Conclusion

In rolling out thousands of almost identical stores across the world, McDonald's has become a familiar landmark and environment not just for Americans. Whether it symbolises home or America or not, McDonald's does symbolise familiarity. And familiarity slices both ways – evoking emotions that alternate between contentment and contempt. Round-the-world travellers' stories draw out a tension between a sense of craving for the familiarity of McDonald's food and space on the one hand and a sense of guilt for eating at these restaurants on the other hand. Their cravings and guilt map McDonald's as a simultaneously homely, local, foreign and global space, thus narrating McDonald's as a paradoxical space contoured by entangled emotions of desire and guilt.

Drawing on meanings of home as familiarity and security, travellers construct McDonald's and similar restaurants as 'homely'. This desire for 'home', then, folds into debates about globalization, symbolised by global icons such as the Golden Arches. The complex emotions that play out in travellers' stories about McDonald's open out onto a more general ambivalence about globalization and the perceived loss of cultural contrast in a homogenising world. Thus, McDonald's becomes an emotional landscape where conflicting desires for familiarity and for difference are played out amidst Big Macs, Maharaja Burgers, french fries and clean bathrooms.

Acknowledgements

I am grateful to the participants of the Emotional Geographies conference for their useful questions on an earlier version of this paper, to Joyce Davidson, Liz Bondi and Mick Smith for their thoughtful comments, and to the round-the-world travellers who participated in this research.

References

Ahmed, Sara. 2000. *Strange Encounters*. London: Routledge.
Augé, Marc. 1995. *Non-Places*. London: Verso.
Barber, Benjamin. 1995. *Jihad vs. McWorld*. New York: Ballantine Books.
Bauman, Zygmunt. 1998. *Globalization: The Human Consequences*. Cambridge: Polity.
Beynon, J. and Dunkerley, D. 2000. 'General Introduction'. pp. 1-38, in *Globalization: The Reader*, ed. J. Beynon and D. Dunkerley. London: The Athlone Press.
Boorstin, Daniel. 1961. *The Image, or What Happened to the American Dream*. London: Weidenfeld and Nicolson.
Brendon, Piers. 1991. *Thomas Cook: 150 Years of Popular Tourism*. London: Secker and Warburg.
Buzard, James. 1993. *The Beaten Track: European Tourism, Literature, and the Ways to Culture, 1800-1918*. Oxford: Clarendon Press.
Canetti, Laura, Bachar, Eytan and Berry, Elliot M. 2002. 'Food and Emotion'. *Behavioural Processes*. 60(2): 157-164.
Coveney, John. 2000. *Food, Morals and Meaning: The Pleasure and Anxiety of Eating*. London and New York: Routledge.
Featherstone, Mike. 1995. *Undoing Culture*. London: Sage.
Fortier, Anne-Marie. 2003. 'Making Home: Queer Migrations and "Motions of Attachment"'. In *Uprootings/Regroundings: Questions of Home and Migration*, ed. S. Ahmed, C. Castañeda, A.-M. Fortier and M. Sheller. Oxford: Berg.
Gupta, Akhil and Ferguson, James. 1992. 'Beyond "Culture": Space, Identity, and the Politics of Difference'. *Cultural Anthropology*. 7 (1): 6-23.

Holmes, David. 2001. 'Virtual Globalization – An Introduction'. pp. 1-53, in *Virtual Globalization: Virtual Spaces/Tourist Spaces*, ed. D. Holmes. London: Routledge.

Klein, Naomi. 2001. *No Logo*. London: Flamingo.

Long, Lucy. Ed. 2003. *Culinary Tourism: Exploring the Other through Food*. Lexington: University of Kentucky Press. Forthcoming.

McDonald's Corporate Website. 2002. http://www.mcdonalds.com (19 September 2002).

Ritzer, George. 1993. *The McDonaldization of Society*. London: Sage.

_____. 2002. *McDonaldization: The Reader*. London: Sage.

Ritzer, George and Liska, Allan. 1997. '"McDisneyization" and "Post-tourism": Complementary Perspectives on Contemporary Tourism'. pp. 96-109, in *Touring Cultures*, ed. C. Rojek and J. Urry. London: Routledge.

Robbins, Bruce. 1999. *Feeling Global*. New York and London: New York University Press.

Robertson, Roland. 1995. 'Glocalization: Time-Space and Homogeneity-Heterogeneity'. pp. 25-44, in *Global Modernities*, ed. M. Featherstone, S. Lash and R. Robertson. London: Sage.

Rojek, Chris. 1998. 'Cybertourism and the Phantasmagoria of Space'. pp. 33-48, in *Destinations*, ed. G. Ringer. London: Routledge.

Schlosser, Eric. 2002. *Fast Food Nation*. London: Penguin.

Sharpley, Richard. 1999. *Tourism, Tourists and Society*, 2nd edition. Cambridgeshire: ELM Publications.

Urry, John. 1990. *The Tourist Gaze*. London: Sage.

_____. 1995. *Consuming Places*. London: Routledge.

_____. 2000. *Sociology Beyond Societies*. London: Routledge.

Watson, James L. 1997. 'Introduction: Transnationalism, Localization, and Fast Foods in East Asia'. pp. 1-38, in *Golden Arches East: McDonald's in East Asia*, ed. J. L. Watson. Stanford: Stanford University Press.

Round-The-World Websites

Ben and Vija's RTW Adventure. 2000. http://benandvija.tripod.com (5 July 2001).

Big Adventures. 2001. http://www.bigadventures.com/global/ (27 June 2002).

Boots 'n' All. 2002. 'Round-the-World Traveller Profiles'. http://www.bootsnall.com (5 August 2002).

Drive Around the World: Latitude. 1999. http://www.aroundtheworld1999.com (17 August 2001).

Marie's World Tour. 2001. http://www.mariesworldtour.com (live chat hosted 28 December 2001; site accessed 14 February 2002).

World Hop. 1998. http://www.worldhop.com (15 July 2001).

World Wander to the World Wonders. 1998. http://www.worldwander.com (19 July 2001).

Chapter 6

The Place of Emotions within Place

John Urry

For a decade or so I have been interested in how it is that visitors (and indeed local people) experience place. What are the pleasures of place? What emotions are provoked by being in a relatively unfamiliar place? How do we learn to release appropriate emotions in those other places? What are the different senses mobilised by being elsewhere? What is involved in 'touring' other places?

I will not deal with all these issues but will develop one theme that relates to shifts in the nature of place. This theme can be captured though a distinction present in Wordsworth between land and landscape as distinct forms of belongingness (Milton 1993). The former conceptualises *land* as a physical, tangible resource that can be ploughed, sown, grazed and built upon. Land is a place of work conceived functionally. As a tangible resource, land is bought and sold, inherited and left to children. To dwell on a farm is to participate in a pattern of life where productive and unproductive activities resonate with each other and with very particular tracts of land, whose history and geography will be known in detail. There is a lack of distance between people and things. Emotions are intimately tied into place, rather as Sarah Hall (2002) describes in her evocative novel *Haweswater* set in the village of Mardale in 1936.

The practice of land is quite different from that of *landscape*. The practice of landscape entails an intangible resource whose definitive feature is a place's appearance or look (Milton 1993). This notion emphasises leisure, relaxation and the visual consumption of place especially by those 'touring'. As Judith Adler (1989) shows there developed in western Europe from the eighteenth century onwards a specialised *visual* sense. This was based upon a variety of novel technologies, the camera obscura, the claude glass, guidebooks, the widespread knowledge of routes, the art of sketching, the balcony, photography and so on (Ousby, 1990). Areas of often wild, barren nature, once sources of terror and fear, were transformed into landscape, what Raymond Williams (1972: 160) terms 'scenery, landscape, image, fresh air', places waiting at a distance for visual consumption by those visiting from 'dark satanic mills' (Macnaghten and Urry, 1998). By 1844 Wordsworth noted that the idea of landscape was a recent development. But within a few years houses were being built with regard to their 'prospects' as though they were a kind of 'camera' (Abercrombie and Longhurst, 1998: 79). The language of views thus prescribed a particular visual structure to the emotional experience of place. Land gave way to landscape (Green, 1990: 88). As

Miss Bartlett paradigmatically declares in *A Room with a View*: 'A view? Oh a view! How delightful a view is!' (Forster 1955: 8, orig 1908).

This transition can be seen in the English Lake District. A place of 'land', according to Daniel Defoe, of inhospitable terror, came to be transformed into 'landscape', a place of beauty, emotion and desire (Urry, 1995). Similarly, the Alps before the end of the eighteenth century had been regarded as mountains of immense inhospitality, ugliness and terror. But they too became 'civilised'. Ring (2000: 9) describes how the Alps 'are not simply the Alps. They are a unique visual, cultural, geological and natural phenomenon, indissolubly wed to European history'. And by the end of the eighteenth century the land of Caribbean 'tropical nature' had been romanticised by European travellers who began to see the scenery as though it were a 'painting', as landscape (Sheller, 2002). And there are countless other examples of how places of land, became places of visual desire, as the inhospitable was turned into a place of emotion, of landscape, especially for rich (male) European visitors.

In this irreversible shift to landscape, the 1840s technology of photography plays a seminal role. Touring and photography could be said to commence in the 'west' around 1840. Louis Daguerre and Fox Talbot announced their somewhat different 'inventions' of the camera, in 1839 and 1840. In 1841, Thomas Cook organised what is now regarded as the first packaged 'tour'; the first railway hotel was opened in York just before the 1840s railway mania; the first national railway timetable, Bradshaws, was published; Cunard started the first ever Ocean steamship service; and Wells Fargo, the forerunner of American Express, began stagecoach services across the American west (Lash and Urry, 1994: 261). This I have argued is the moment when the 'tourist gaze' emerges, involving the combining together of the means of collective travel, the desire for travel, the techniques of visual reproduction and the emotion of landscape (Urry, 2002). As a visitor to Victoria Falls subsequently declared: 'Wow, that's so postcard' (quoted Osborne 2000: 79) as landscape rather than land had become all the rage; with even Ruskin declaring that daguerreotypes are 'glorious things ... nearly the same as carrying off a palace itself' (quoted Botton, de 2002: 223).

Moreover, one particular way of experiencing place became particularly valued, what I have called the *romantic* gaze. In this what is emphasised is a solitudinous, personal, semi-spiritual relationship with place. People expect to experience the place privately or at least only with 'significant others'. Large numbers of other visitors, as at the Taj Mahal, intrude upon and spoil that lonely contemplation desired by western visitors, as famously seen in the Princess Diana shot at the Taj (Edensor, 1998: 121-3). The romantic gaze involves further quests for ever new objects of this solitary experience, as reflected in Alex Garland's *The Beach* (1997), a process like the sorcerer's apprentice, consuming and devouring the very places that are sought out for the emotional and solitary appropriation of place.

By contrast what I have called the *collective* tourist gaze involves conviviality. Other people also in that place give liveliness or a sense of carnival or movement. Large numbers of people that are present indicate that this is the place to be. These moving, viewing others are obligatory for the emotional experience of place, as in a cosmopolitan New York, on a beach in Rio, in the casinos of Las Vegas, at the Sydney Olympics, in a club in Ibiza and so on. Baudelaire's account of flânerie captures this emotional immersion; he describes: 'dwelling in the throng, in the ebb and flo, the bustle, the fleeting' (cited by Tester, 1994: 2).

I have so far talked about the pleasures of place, noting the shift from land to landscape. I also distinguished between the romantic and collective forms of landscape. However, the notion of place needs to be further developed here.

Place should not be thought of as an abstract Cartesian space that can be defined by various geometric coordinates. Rather places are centres of many material activities, including the purchase and use of goods and services. And very many places across the globe are being restructured as places of consumption, of what Fainstein and Judd (1999) term 'places to play'. Places are emotionally pleasurable because they are sites of intense and heightened consumption, locations within which distinct goods and services are compared, evaluated, purchased and used (and over-used). Places to play are often places of excess, where consumption is taken to extreme. Examples of such consumption taken to excess include gambling in Las Vegas, Broadway shows in New York, extreme sports in Queenstown, New Zealand, country house meals in the English Lake District, exotic sex tourism in Thailand, water sports in the Caribbean, recreational drugs in Ibiza, whisky in Scotland and so on.

The pleasure of such places derives from the consumption of goods and services that somehow stand for or signify that place. Through consuming certain goods and services the place itself comes to be experienced. The good or service is metonymic of the place, with the part standing for the whole. The consuming of place involves the consumption of goods/services that are somehow unique or at least culturally specific to that place. People eat, drink, gamble, waterski, smoke, bungee jump, the 'other' (see Urry 2002: 3).

Or so people hope. But often of course places are places of disappointment, frustration, bitterness, perhaps best captured in *The Beach* (Garland, 1997). There is often a massive gap between what people anticipate will be a place's pleasures and what is actually encountered. Thus the items of consumption may not be available (the hotel is closed for the winter), or the services have become too commercial (as on the beach at Goa), or the service delivery has become too expensive (as in a Parisian restaurant) or too low quality (as in a guest house in a fading British seaside resort such as Morecambe), or the arts shops have turned into souvenir stalls (as at Albert Dock in London) and so on.

In many ways the pleasures of place are thus contradicted by the actual consumption possibilities, especially with the domination of the world economy with huge homogenising capitalist corporations. Such companies often fail to ensure the specificity of the commodity, or of attracting other consumers consistent with the emotional pleasures of that place. So emotionally experiencing a place through consuming certain goods and services is shot through with contradiction and ambiguity (see Urry 2002 and also Chapters 4, 5 and 9 in this volume).

The dynamic conception of place can be further explored through Kevin Hetherington's (1997: 185-9) notion of place as a 'place of movement'. 'Imagine', he says, 'place as being like a ship' (Hetherington, 1997: 185). They are not something that stays in one location but move around within networks of agents, humans and non-humans. Places are about relationships, about the placings of materials and the system of difference that they perform. Places are located in relation to sets of objects rather than being fixed only through subjects and their uniquely human meanings and interactions.

I take three points from this analysis. First, objects are highly significant in the nature of place. Various objects constitute the basis of an 'imagined presence', carrying that imagined presence across the members of a local community. Places also carry traces of the memories of different social groups who have lived in or passed through that place. All sorts of contestation over those memories mark off each place. Various objects can function in this way and not just the immense, official monuments of community. We might also note the incredible significance of various kinds of buildings as central to place. The building in Bilbao of the Frank Gehry Guggenheim is a classic example of the power of landmark building and celebrity-architects to reposition place, to move Bilbao closer to the global centre.

Second, places can be distinguished in terms of whether they are temporally rich or poor. Richard Sennett (1991: Chapter 7) for example says that some 'places [are] full of time' and it is this that makes them brim with 'cosmopolitan opportunity'. They are based upon instantaneous time, a time induced by the dazzling disorientation of Virilio's 'speed' (1986). Other places exhibit a 'drudgery of place', the sense of being inexorably tied there and where time seems fixed and unchanging. Such places remain heavy with time. Some places are thus left behind in the 'slow lane', as with many English seaside resorts.

And third, places even based upon a high degree of geographical propinquity depend upon movement. Paths can show the accumulated imprint of countless journeys that have been made, as people go about their day-to-day business. The network of paths shows the sedimented activity of a community stretching over many generations; it is what Tim Ingold (1993b: 167) terms the taskscape made visible (Macnaghten and Urry 1998). People imagine themselves treading the same paths as countless earlier generations that have lived there or thereabouts.

But also places are massively interconnected to many other places through movement. Raymond Williams (1988) in the novel *Border Country* is, according to Tony Pinkney (1991: 49), 'fascinated by the networks men and women set up, the trails and territorial structures they make as they move across a region, and the ways these interact or interfere with each other'. Likewise Henry Thoreau (1927: 103) in his evocative return to 'nature' on the banks of Walden Pond in the mid-nineteenth century did not complain about the sound of the railway. He considered that he was: 'refreshed and expanded when the freight train rattles past me, and I smell the stores which go dispensing their odours all the way from Long Wharf to Lake Champlain, reminding me of foreign parts ... and the extent of the globe. I feel more like a citizen of the world'.

This movement of place can be seen in the history of the Lake District (see Urry 1995: Chapter 13). This place in the north west only really became part of England when many visitors, especially artists and writers, travelled to it from the metropolitan centre at the end of the eighteenth century onwards. These visitors, with their poetic reassessment in terms of the picturesque and the sublime of the objects of mountains, lakes, tarns and waterfalls, moved the Lake counties closer to the centre of England. Land got changed into landscape through artists and writers 'moving' the Lake counties into English culture. It had previously been 'on the margins', left behind in the slow lane of eighteenth century English life.

Many of the key writers were deemed to be from that place, and became known as the 'Lake poets' (whether or not they were 'local'). The Wordsworths, Southey, Coleridge and so on became celebrities in an area previously without national celebrities. They became major tourist attractions especially for metropolitan visitors. By the 1840s Wordsworth was receiving 500 visitors a year at Rydal Mount. And after their death the Lake poets were transformed into literary shrines and memorialised as core figures at the very heart of English literature. These visitors had brought the peripheral and background area of inhospitable terror, of land, closer to the centre, almost part of a metropolitan nature. And this parallels the process that Nicholas Green (1990: 88) describes as the 'metropolitanising of nature' around Paris in the mid-nineteenth century.

But this was achieved at a cost. For the emotional pleasures of place to be achieved through this moving to the 'centre' involves further shifts from land to landscape. E.M. Forster in *Howard's End* characterises the process by which certain places, like London or Paris, have come to be nomadic or cosmopolitan. He argued that 'Under cosmopolitanism ... we shall receive no help from the earth. Trees and meadows and mountains will only be a spectacle ...' or what I have called above 'landscape' (E.M. Forster 1931: 243; see Szerszynski and Urry 2002, for research related to the following). Certain places seem quintessentially cosmopolitan and this is what makes them pleasurable; other places are not. And certain sorts of places come to be detached from nature and the physical environment. Nature is transformed into landscape, comprised of images of trees, meadows and mountains that are to be known about, compared, evaluated, possessed, but not places that can be 'dwelt-within' as land.

It seems that, as visuality has become central to the experience of place, so it has turned into an abstracted, disembodied quality or capacity. There is thus a tendency for *all* places in the end to become cosmopolitan and nomadic. The related shift to a *visual* economy of nature – the assumption that nature and place are above all to be looked at rather than used and appropriated – assists this 'de-substantialisation' of place. A given locality becomes not a unique place, with its own associations and meanings for those dwelling or even visiting there, but a particular combination of abstract characteristics, which mark it out as similar or different, as more or less scenic than other places.

The language of landscape is thus a language of mobility, of abstract characteristics. It is not just that such mobility is necessary if one is to develop the capacity to be reflexive about landscape. It is also that landscape talk is itself an expression of the life-world of mobile groups, as Bron Szerszynski and I have shown elsewhere (2002). These mobile groups include both tourists and environmentalists as Buzard (1993) brings out well. They also include elderly women from the Isle of Skye as Sharon Macdonald (1997) illustrates in recounting a common story heard, which runs as follows:

There was an old woman ... living in township X. One day a couple of tourists come
by and start asking her questions.
'Have you ever been outside this village?' ...
'Well, yes. I was at my sisters in [neighbouring township] ...
'But you've never been off the island?'
'Well, I have, though not often I suppose'.
'So, you've been to the mainland?' She nods. 'So you found Inverness a big city then?
'Well, not so big as Paris, New York or Sydney, of course ...'

Thus it seems that almost all places are 'toured' and the pleasures of place
derive at least in part from the emotions involved in visual consumption of place.
This produces the emotion of movement, of bodies, images, information, moving
over, under and across the globe and reflexively monitoring places in terms of
abstract characteristics. Those mobilities, a 'fluid modernity' according to Zygmunt
Bauman (2000), have produced a widespread capacity for aesthetic judgment that in
turn feeds into and animates global tourism as well as the environmental movement.

And this is judgment from afar, not necessarily 'grounded', a judgment
possessive and abstract. In *The Beaten Track* Buzard notes how Wordsworth's *The
Brother* 'signifies the beginning of modernity ... a time when one stops belonging
to a culture and can only tour it' (Buzard 1993: 27). Thus our destiny is to find
pleasure in place through an unrelenting visual economy of signs although of
course diverse other senses get mobilised at the margins, as occasional resistances,
as resistant *Bodies of Nature* (Macnaghten and Urry 2001).

As Alain de Botton (2002: 223) notes 'Technology may make it easier to
reach beauty, but it has not simplified the process of possessing or appreciating it'.
Places have thus turned into a set of abstract characteristics in a mobile world, ever
easier to get to, but not appreciated from within. The tourism industry rushes
headlong to search for new 'rooms with a view' before they are 'postcarded'. And
we are all consumed in this. Experiencing place as landscape, as something to tour,
is our destiny. It cannot be avoided and the emotions we experience as poets,
novelists, older women on the Isle of Skye, environmentalists or tourists are all
judgments from afar, abstract and mobile.

And this somewhat paradoxically parallels the judgements made by the
Manchester Water Company when it sized up the building of the dam in Haweswater
and submerged the land of Mardale in the late 1930s. A new landscape was born as
the land disappeared under the dark torrent of water rushing into the valley and
forming a modern reservoir, as Sarah Hall's (2002) novel evocatively recounts.
Valleys get submerged under floods of water or of tourists or of environmentalists or
even of locals who are all destined only to enjoy the emotions of 'touring'.

References

Abercrombie, N. and Longhurst, B. 1998. *Audiences*. London: Sage.
Bauman, Z. 2000. *Liquid Modernity*. Cambridge: Polity.
Botton, A. de 2002. *The Art of Travel*. London: Hamish Hamilton.
Buzard, J. 1993. *The Beaten Track*. Oxford: Clarendon Press.

Dicks, B. 2000. *Heritage, Place and Community*. Cardiff: University of Wales Press.

Edensor, T. 1998. *Tourists at the Taj*. London: Routledge.

Fainstein, S. and Judd, D. (eds) 1999. *The Tourist City*. Cornell: Yale University Press.

Forster, E.M. [1910] 1931. *Howard's End*. Harmondsworth: Penguin.

Forster, E.M. [1908]1955. *A Room with a View*. Harmondsworth: Penguin.

Garland, A. 1997. *The Beach*. Harmondsworth: Penguin.

Green, N. 1990. *The Spectacle of Nature*. Manchester: Manchester University.

Hall, S. 2002. *Haweswater*. London: Faber and Faber.

Hetherington, K. 1997. 'In place of geometry: the materiality of place', in K. Hetherington and R. Munro (eds) *Ideas of Difference*. Oxford: Blackwell/Sociological Review. pp. 183-199.

Lash, S. and Urry, J. 1994. *Economies of Signs and Space*. London: Sage.

Macdonald, S. 1997. 'A people's story: heritage, identity and authenticity', in C. Rojek and J. Urry (eds) *Touring Cultures*. London: Routledge. pp. 155-75.

Macnaghten, P. and Urry, J. 1998. *Contested Natures*. London: Sage.

Macnaghten, P. and Urry, J. (eds) 2001. *Bodies of nature*. London: Sage.

Milton, K. 1993. 'Land or landscape: rural planning policy and the symbolic construction of the countryside', in M. Murray and J. Greer (eds) *Rural development in Ireland*. Aldershot: Avebury.

Osborne, P. 2000. *Travelling Light. Photography, travel and visual culture*. Manchester: Manchester University Press.

Ousby, I. 1990. *The Englishman's England*. Cambridge: Cambridge University Press.

Pinkney, T. 1991. *Raymond Williams*. Bridgend: Seren Books.

Ring, J. 2000. *How the English Made the Alps*. London: John Murray.

Samuel, R. 1994. *Theatres of Memory*. Verso, London.

Sheller, M. 2002. *Consuming the Caribbean*. London: Routledge.

Szerszynski, B. and Urry, J. 2002. 'Cultures of cosmopolitanism', *Sociological Review*, 50: 461-81.

Tester, K. (ed) 1994. *The Flâneur*. London: Routledge.

Urry, J. 1995. *Consuming Places*. London: Routledge.

Urry, J. 2002. *The Tourist Gaze. Second Edition*. London: Sage.

Virilio, P. 1986. *Speed and Politics*. New York: Semiotext(e).

Williams, R. 1972. 'Ideas of nature', in J. Benthall (ed) *Ecology. The Shaping Enquiry*. London: Longman.

Williams, R. 1988. *Border Country*. London: Hogarth Press.

SECTION TWO
RELATING EMOTION

Chapter 7

'Not a Display of Emotions': Emotional Geographies in the Scottish Highlands

Hester Parr, Chris Philo and Nicola Burns

Introduction

> It was many years before I could walk through the street and hold my head up and feel that, you know, ... I belonged. ... [F]or many years, I felt ... I felt that I didn't belong here, you know, I didn't feel like I was part of the community. [Member of a discussion group consisting of people experiencing mental health problems and living in the Skye and Lochalsh area]

This chapter is concerned with the emotional geographies of people with mental health problems living in the rural and remote places of the Scottish Highlands. It draws on empirical materials from a qualitative research project with over 100 users of mental health services (see Parr *et al.* 2004). This work has repeatedly highlighted emotional issues: why and how some interviewees feel supported, included and cared for, and how others feel rejected, excluded and isolated (as in the above quotation). We have encountered a diversity of emotional states, bound into a variety of relations, practices and exchanges, some of which reflect specific 'illness' experiences, but others of which are clearly framed by the micro-dynamics of community life in the places concerned.

Notwithstanding important variations across the Highlands, certain commonalities have emerged from our interviews that suggest we can speak of the region as a distinctive 'emotional terrain', marked by widely shared beliefs about the inappropriateness of disruptive emotional display. Such beliefs, we argue, serve to 'discipline' people with mental health problems, forcing them to conceal their difficulties in ways that might ultimately be unhealthy for them. Such strategies are the result of complicated linkages between individuals and communities, and between individual feelings and collective beliefs about emotional conduct in this part of the world. There are numerous 'emotional geographies' in play here: ones to do with the community dynamics of emotional (non)display in everyday social spaces, but also ones to do with largely taken-for-granted imaginings of regional history, culture, identity and behaviour. There are various ways in which we could probe these geographies (see also Parr *et al.* 2003a, 2003b, 2004), but as one thread in what follows we explore, tentatively and informed by empirical findings, the notion of 'repression'. We begin by reviewing how this notion has been mobilised before in research concerned with the Highlands, notably in studies that draw upon Freudian concepts of repression. While having reservations about such studies, they do highlight the possibilities for a 'socialised' and 'placed' approach to thinking through the enactment of repression. In the empirical heart of the chapter, we seek to show the utility of such an approach, underlining the extent to which the emotional terrain of the Highlands is layered by silence, silencing and self-silencing. Because the unspeakability of emotions becomes a key theme of the empirical evidence, we then allow a shift in conceptual register in conclusion to inspect the claims of non-representational theory. We conclude by stressing that emotional (non-)display is best understood *not* as due to some de-socialised and placeless process, whether this be construed as psycho-dynamic repression or as some ultimate failure of representation; rather, for us, emotions must be seen as unavoidably situated, in this case in the rural Highlands.

Repression, Mental Health and the Scottish Highlands

In discussing the problematic imaginative geographies of Hebridian[1] Scottishness in the context of his ethnography of Presbyterianism, McDonald (2000, p. 7) argues that recent scholarship on the subject has only served to 'perpetuate exotic mythologies of Calvinism (alcoholism, mental illness, gloominess) while being simultaneously blind to its everyday meaning and politics'. For McDonald (p. 8), these mythologies rest on readings of an 'internally repressed' religious community which pay little attention to the complicated social spaces and politics of integration and belonging actually experienced by church members. In his critique of Agnew's work, for example, he disputes that the influence of Presbyterian ministers 'undoubtedly contributes to a number of social problems, such as high levels of anxiety and obsessional disorders among Hebridian women and very high levels of alcoholism among the menfolk' (Agnew, 1996, p. 36). McDonald's critique of scholarly assumptions about Hebridian religious life and its influence on mental health is clearly instructive for us in that he is keen to distance himself from blanket readings of Hebridian emotionality; we will return to this point below.

Thinking beyond Hebridian Presbyterianism, assumptions about a regional unhappiness and an 'internally repressed' people are applied to the wider Highlands by a range of commentators (Free Church of Scotland, 1998; Hunter, 1985; Macritchie, 1994; McDonald, 1994). The mental and emotional state of Highlanders is clearly of concern and curiosity in such writings. In Parman's rich, although controversial, ethnography of Highland culture, she traces the significance of emotional display in the regional context. She claims that displayed 'anxiety' amongst Highlanders is culturally significant, often read by religious followers as a possible sign of *Curam* (Gaelic for 'conversion to faith'). However, and importantly, if this state is not accompanied by obvious religious conversion, then it is likely that those experiencing visible anxiety 'would have ended up in Craig Dunain (the mental hospital near Inverness)' (Parman, 1990, p. 148). In both cases, displayed anxiety is read as likely to result in forms of institutionalisation (within the church or the hospital).

More scientifically orientated studies (Brown and Harris, 1978; Brown *et al.*, 1977; Brown and Prudo, 1981; Prudo *et al.*, 1981; Prudo *et al.*, 1984) have postulated that there are high rates of anxiety among women living in remote parts of rural Scotland such as the Outer Hebrides. Such studies have sought to compare rates of depression and anxiety amongst women in an inner-city community in London and communities on the Isles of Lewis and Uist in the Outer Hebrides. Their findings suggest a complicated picture of psychological and emotional life in these latter places, with rare signs of depression amongst island women considered to be highly integrated into local communities. However, these women *do* appear to show relatively high levels of 'anxiety' and 'obsessional disorders'. Brown *et al.* (1977, p. 374) speculate that strong social forces in rural island communities create distinctive normative frameworks that lead to such anxious behaviour: 'The strength of such norms may well create conflicts whose repression (in a Freudian sense) leads to a risk of anxiety states'. These authors go further, suggesting that such 'anxiety may be a cultural characteristic of island women', a state connected in part to proximate patriarchal networks in which fear of community ostracism (amongst other things) dominate senses of ontological security. It is of curious significance, given Parman's (1990) claim that while anxiety is considered highly prevalent in the Highlands, the social consequences of visible anxious emotions are not discussed by the above studies.

The above readings of remote rural Scottish emotional life are striking in their similarities: it seems that anxiety, unhappiness and particularly 'repression' are strong features of the sociality of the Highlands and Islands. We would like to pick up on one aspect of this problematic picturing of rural Highland life and discuss it further. 'Repression' – which in the hands of Brown, co-workers and others is acknowledged to have Freudian roots – is arguably under-theorised and under-investigated as a feature of everyday geographies, and it may be that here, as elsewhere, the borrowings from Freud within psychoanalytically informed geographical research are as yet limited (Callard, 2003; Philo *et al.*, 2003). Although several writers above mention repression, they are less than analytically clear about its significance for interpreting the Highlands. Repression can be understood as involving a 'process (defence mechanism) by which an unacceptable impulse or idea is rendered unconscious' (Rycroft, 1995, p. 157). Classic Freudian understandings of repression emphasise how 'the essence of repression lies simply in turning something away, and keeping it at a distance from the conscious' (Freud, 1915, p. 147; cited in both Badcock, 1988, p. 15, and Erwin, 2002, p. 478). Repression 'stands in place of painful conflict', writes Badcock (1988, p. 17), and it entails the individual's psychic economy attempting 'to escape pain' so that we are insulated from anything troubling or anxiety-inducing.[2] As Pile (1996, pp. 109-111) argues when introducing psychoanalysis to geographers, the Freudian conception of repression relies on the notion that from early childhood we learn to deny and to deflect sexual drives and desires. The emotional energy of this repressed desire is then potentially available to be redirected elsewhere, such as into neuroses, once we become adults. 'The unpleasure ... released by the liberation of the repressed' (Freud, 1947, p. 15; in Badcock, 1988, p. 17) is thus reckoned to have the power to induce 'pathological' mental states, and in this vision it is actually the *failure* of repression that risks someone slipping into being mentally unwell.

Reading the Highlands and its mental health stories through such ideas is problematic, partly because it picks up on seemingly individualised understandings of the self and the unconscious. These understandings arguably risk misrepresenting the contextual quality of psychoanalytic thinking, evident from Freud's account of social and cultural life in his work on groups and civilisations (Bondi, 2004, pers comm), but some accounts of repression do still position this phenomenon as about little more than the working of an individual's 'lonely' unconscious. Repression 'occurs only within the mind; and only the individual in question can be responsible for it', states Badcock (1988, p. 18), implying that social processes are unimportant in prompting people to strive, perhaps initially quite consciously, to cordon off their anxieties, drives and wishes from the realm of everyday interaction. Somewhat differently, Brown *et al.* (1977), influenced by Freudian ideas, in effect 'socialise' repression for their Highland research, principally by invoking Durkheim's concept of fatalism in order to express how aspects of Highland society involve 'oppressive' relations which result in psycho-social 'blocking'. Here fatalism involves:

> ... an excess of regulation found in persons whose future is relentlessly blocked, whose passions are violently *repressed* by an oppressive discipline, by the unavoidable and inflexible roles over which one is powerless. (Durkheim, 1952, p. 63; cited in Brown *et al.*, 1977, p. 374)

These comments, discussed in the context of 'punitive' features of island life for Hebridian women and variously related to strict behavioural codes and fear of gossipy disapproval, offer an attempt to enrol a version of the repression concept into an explanation of thoroughly *placed* anxious emotional sociality. By 'placed', we mean that the psychodynamics involved are not anchored solely in the individual, but rather in a more collective, shared, socially, culturally and even environmentally situated play of exchanges – we would say indelibly emotional exchanges – leading some vulnerable individuals to conceal certain aspects of their emotionality. In this account the initial *making* of repression is thoroughly woven with the relational realities of everyday emotional life, and is therefore something that varies in intensity and effect from one place to another, creating the possibility of particular regional (or other such scaled) emotional geographies.

Brown *et al.'s* (1977) study and the others associated with it (see above) are primarily quantitative psychiatric surveys that, only by way of speculative discussion rather than through grounded qualitative research, assert such ideas. This should not detract from the importance of these attempts, however, nor from the interesting potential of trying to harness the repression concept into a more social-cultural geographical framework. Our interest and intervention is arguably subtler, though, in that we wish to explore how emotional and psychological 'disruptions' are subject to repressive relations, ones both individually and collectively constructed. In other words, we consider how mental health problems (including anxieties, depressions and other emotional and psychological states) are commonly *denied* in Highland places both by people who experience them and by people who witness them in others. Nonwithstanding McDonald's (2000, p. 7) fears about echoing crude signifiers of repression as aired in uncritical Highland mythologies, we wish to rise to the challenge of providing materials that illuminate the 'everyday meaning and politics' of repression, as bound into community encounters involving people with mental health problems. In tracing aspects of the experienced emotions of Highlanders living with mental health problems, as well as their attempts to deny emotional trauma, we effectively comment on both individual *and* collective repressive practices, albeit as occurring in a largely conscious realm. In so doing, we seek to 'socialise' and to 'place' repression in an endeavour to understand repressive practices as integral to the human geography of the remote rural Highlands.

A Geography of Highland Emotionality and Mental Health Problems

The Highlands are often associated with a romantic, idealised image of rural community life characterised by closeness, intimacy and a certain kind of hardy spirit. As elaborated below, our interviewees provide a more mundane picture of an emotional sociality that impacts quite directly on their experience of living with mental health problems in remote places. In particular, they convey how difficult it is to live with mental health problems in a rural region with insufficient people or centres of population to be anonymous, as they commonly envisage, one might in an urban centre, and this simple assumption lies at the heart of everything that follows (see Parr *et al.*, 2004; Philo *et al.*, 2003). Our comments draw mostly on interviews with people experiencing mental health problems in remote parts of Skye and Lochalsh (SL), Easter Ross (ER) and North-West Sutherland (NWS), crofting landscapes with sparse, scattered populations. In these areas formal mental health services are few and amount to a handful of community psychiatric nurses who visit people in their own homes and travel across huge geographical areas on a daily basis (Parr *et al.*, 2003b). There are therefore very few services available to help with traumatic mental health issues, and this means few outlets within which individuals can speak openly about mental health difficulties. As a result sufferers have little choice but to fall back on the communicative resources of family and other local people, but, as we see below, such a solution is far from straightforward. There also exists a complicated and nuanced politics associated with the disclosure and discussion of emotional and psychological difficulties in small communities. It is our contention that these complications and their consequences involve particular geographies, critical appraisal of which can shed light on the remote rural Highlands as a distinctive (if hardly unique) emotional terrain.

Cultures of resilience, cultures of silence and mental health problems

> It is a different culture up here, a different way of life. [Gill, SL]

Accepting that our assertions here are partial and not unproblematic, we wish to venture a claim that remote rural Highland life is characterised by particular cultures of emotional exchange, which impact on people with mental health problems in distinct ways. The first thing to note is that this culture, according to interviews, is very much a product of historical circumstances, which many now admit is being diluted by the realities of 21st century technologies, media influences, population influx and so on. A sense of this culture still persists, however, especially for older Highland residents who routinely draw on the traumatic histories of 'The Clearances' and the sense of collective resilience that has resulted from this period to explain their emotional place-in-the-world as sorrowfully 'stoic' (Hunter, 1985). This shared sense of history means that 'There is a kinship amongst Highland folk' (Chloe, SL), a kinship which readily recognises an imagined and shared emotional past:

> The type of life that people had here was so hard that the result of it … [was people] not willing to trust or be emotional to anyone, or invite anyone in. That generation was just surviving, they didn't have the time or inclination to get to know anybody outside their immediate family. [Roisin, ER]

In addition to claims about cross-generational emotional traits in the region, many interviewees referenced the role of religion as an important influence in their thinking about and practice of emotional relations. One interviewee, a church elder, when prompted about the role of the church in talking about people's emotional states simply replied that '[w]e don't mention emotion at all' [Darren, NWS]. Another interviewee provided a detailed assessment of how this attitude can lead to a denial of emotions:

> On Skye, you'll find this … Presbyterianism that will not allow you to show your affections, so people clam up. … My wife was … she was taught you don't show your affections, you don't hold your husband's hand in public, you don't put your arm through his in public … [Alistair, SL]

Such embodied emotional restrictions engendered through the strong influence of the church and pervading through community life dominated many interviewees' explanations of cultural influences on the problematic disclosure of mental health problems, the experience of which involves disruptive emotions. Such historical and religious influences may evoke stereotypical notions of a regional 'resilience', 'stoicism' and moral strictness, as McDonald (2000) worries, but when challenged with such a view, interviewees *insisted* that such constructions should remain central to explaining Highland attitudes to mental health problems.

Religious influences aside, there also seemed to be a shared sense about how people in the Highlands should emotionally conduct themselves. A strong notion of there being set rules or 'ways' was conveyed by many of our interviewees: 'it's just the way I think these things are done up here' [Conner, NWS]. Many interviewees appeared acutely conscious of such prevailing rules, and were aware of how they needed to modify the effects of their mental health problems to fit in with these cultural expectations. In particular, there was the feeling that Highlanders *should* be able to cope with whatever life has to throw at them; and, indeed, should be able to cope without 'fuss' or 'drama', and certainly without making any kind of emotional scene:

> In the Highlands … you have to be reliant on yourself. You know everybody has to pull together, so there's no room for people being ill. So it's very much, 'pull yourself together and get on with it'. [Miriam, ER]

Such words transmit the ethos pervading the culture of resilience that interviewees claim to characterise Highland life, and in relation to which mental health problems are antithetical. Any voicing of such problems, any expression of an emotional content that betrays someone *not* 'pulling themselves together', risks over-stepping the lines of Highland culture:

> Well, I think, like, the way a lot of people have been brought up where you don't go out and complain, you know, you don't moan about anything, you keep it to yourself and you just plod on. I think that's a bad thing in a way and that's how a lot of people end up with mental health problems 'cos they can't express themselves when they start to feel ill, you know, they try and battle on and hide it, keep it inside, you know. And then they end up just cracking up, you know. [Ishbel, ER]

The consequences of transgressing cultural norms surrounding emotional expression can include community rejection, devastating in its social impact in small rural places. Hence the individual and collective need to 'keep it inside', as Ishbel says, contributes to the routine avoidance of moments and situations which might reveal or release deeply held trauma, terror, desire or need. In this sense interviewees can be argued to be discussing the socialisation of repression.

Interviewees also listed a host of ways in which they have learned to create emotional distance between their family, friends and neighbours, including not directing questions at people where answers or discussion with a deep emotional resonance might be required:

> Nobody really talks to each other about private things like that. [Conner, NWS]

A key claim of our interviewees is clearly that communal silence exists around problematic and strong emotions, and that as such there is a need constantly to reign in one's own emotional states:

> Not a display of emotions, that's it, that's the word I am looking for. [Pauline, SL]

> Yes … that could be a problem amongst a lot of people here. They don't like to see strong emotions. [Deborah, NWS]

There is much more that we could say about the cultural context to emotional expression in the rural Highlands, but overall what it has meant for Highlanders with mental health problems is a huge sense of social pressure not to disclose or discuss their problematic emotional and psychological states. There is also pressure to battle constantly with the experience of emotional and psychological disruption, not only in terms of searching for recovery from illness, but in respect of the need to conceal what they have learned is inappropriate behaviour. It is to some geographical implications of such cultural norms, and the psycho-social struggles that they produce, which we now turn.

Geographies of Emotional Containment: The Tyranny of Repressive 'Normality'

We can see that due to common perceptions of what we have called a culture of silence and resilience about various emotional and psychological experiences, people with mental health problems are hardly encouraged to seek support and help from their local communities, friends and even family. While this results in feelings of isolation amongst many people in rural Highland places, there are also other consequences, especially in regard to social relations in public spaces. Many interviewees discussed their strategies for occupying everyday public social spaces when experiencing mental health problems. Again, although there were many dimensions to these discussions, that there was a stated need for detailed coping strategies was evidently related to the distinctive emotional content of rural Highland life. Barry articulated why the rural Highlands can be seen as a nuanced geography of noticeable social actions that effectively demands such strategies:

> It's easier to read everything when you're not in the city. Because in the city you have got everything coming at you, you've got neon lights, you've got the noise of the cars, traffic lights, you've got the whole bloody city at you. That really doesn't leave lots of space in your mind for the little nuances of how a person is feeling. People are subtle [in rural Highland places], very good at concealing what they are feeling'. [Barry, SL]

Barry highlights one of the curious contradictions of rural Highland sociality: that whilst people claim that emotions are ignored, buried and contained in everyday social relations, they are also noticed. At least this was a perception translated in many interview accounts, that 'little nuances' of feelings are routinely 'read' by other actors in rural public spaces. Hence for people with mental health issues, there was a need to be continually monitoring their own behaviour and their emotional expression so that they might be 'read' as 'normal' and therefore not risking transgression, community rejection or stigma. As a result of this social 'fine tuning', rural Highlanders become adept at both hiding emotion and concealing their problems in public spaces:

> It's like you have to ... not pretend, that's not the right word, but it's the only word I can think of ... put up a front. It just feels that you're, it's like a goldfish bowl, so people can watch you, people [are] watching ... [Judith, Inverness, the main urban centre and only large town in the Scottish Highlands]

As part of a fear of showing their emotions and of possibly breaking down in public space, interviewees related how they used strategies for deflecting attention during social interactions. As Karen elaborated:

I think what I found was I would maybe meet somebody. If someone just said 'how are you today'? I would say 'fine' and then would go on talking about the weather and I would say 'how are you doing'? I was very aware that I was putting it back to them to get them to talk because I didn't want to. [Karen, ER]

Curiously, then, despite the fear of emotional incompetence in rural public spaces, some interviewees proved themselves to be skilful managers of conversational exchange, in ways that allowed them to conceal their true feelings. This may sound *un*dramatic, but when experiencing acute or chronic depression or feeling the devastating personal effects of severe social anxiety, for example, interacting 'normally' in rural public spaces *through which only a handful of very familiar people pass* is a stunning and quite exhausting achievement.

Not only were interviewees skilful managers, but also effective readers of what particular social interactions demanded of them. As Jessica put it, 'I'm keeping with the rules' [NWS], implying an ability to read and then to react to emotional norms. However, the constant and fearful monitoring of social and emotional boundaries is highly draining, especially if experiencing acute phases of illness, and interviewees exercised strategic spatial avoidance tactics if they felt that their attempts at 'passing' might not be robust enough on any particular day. As Ishbel revealed, 'I take all the short cuts to get my shopping and take all the short cuts back home so I don't have to meet anyone' [Ishbel, ER]. These performances can perhaps be seen as important individual attempts to 'turn away' disruptive emotions from relational exchange in order to maintain bounded senses of self. What these attempts also entail, however, are practices that respond to and subsequently reinforce collective repressive urges to distance culturally unacceptable emotional display from everyday rural spaces, ones particularly marked by social proximity, religious morality, familiarity and 'watchfulness'.

Geographies of public emotional restraint were supplemented for some by private spaces of emotional release, with the home often being cited as a haven where true feelings could be expressed. However, also common amongst some interviewees was even a perceived need to pass as emotionally and psychologically 'normal' in private space, where performing wellness to family and friends was deemed necessary to prevent them feeling guilty about the burden of care that might fall on the shoulders of others:

My children ... you can't be ... I try really hard not to, not [show depression] because I don't want them growing up with a depressed mother. [Julia, NWS]

Here Brown *et al.'s* (1977) speculations about the anxieties of Highland women (see also Thien, this volume and 2005) have a particular resonance with respect to the above comments. In a region where discourses, roles and practices relating to strongly domesticated femininities are prevalent, home-space is not necessary a significant geography of emotional release, since disruptive emotions here may be understood as signs of weakness (see also Bell and Valentine 1995). Compounding these processes of self-restraint for such women is the knowledge that other family members are acculturated into regional norms of emotional denial:

It exists even in my own household, because my husband does not understand emotional problems at all ... I've learnt the best thing is not to say ... I can't even say in this house 'Oh God I'm depressed' ... So you learn to hide it. And maybe that way you control it. I don't know ... My husband I think denies himself lots of things ... he won't give in to emotion ... he has a horror of emotional stuff. If I were to burst into tears, he'd run a mile. [Deborah, NWS]

Not 'showing', 'saying' or obviously 'embodying' strongly negative emotions results in highly competent emotional management strategies through different spaces.

This competency nonetheless belies how difficult denial, distancing, passing and hiding can be, and how utterly exhausting are the efforts at understanding and acting emotional and behavioural normality. Julia shows how this painful process is characterised by a constant struggle to articulate oneself coherently in the world:

It's probably not so much concealing as trying to behave normally ... trying to be ... You feel dreadful ... you feel as if you are not functioning properly but you can't describe it, you can't discuss it, you can't just be, you just feel so awful that you're not ... You try to be ... You try to live with it. [Julia, SL]

Underlying much of the struggle to hide problematic emotions and to pass as 'normal' is the fear of the social consequences of not passing, of being the subject of stares, observation and gossip, and the perceived rejection that might follow. As Fred says 'I've got to act normal ... I can't show [the illness] ... then people are going to back away again' [Fred, NWS]. In this chapter we have not documented in detail what happens when repressive strategies fail on an individual and collective basis, as they often do for people with severe and enduring mental health problems. For some, it certainly means rejection, isolation and exclusion from previously supportive and dense community networks, while for others it means entry into more open and caring relationships that are effected through both informal and formal pathways. There is no one clear-cut Highland story of what happens when mental health problems are *not* concealed (Parr *et al.*, 2004). Rather, it is the *perception* of transgression and collapse that defines this regional geography of emotional sociality.

Conclusion: Representing Emotional Terrains?

We have sought to lay out some ingredients for an analysis of a distinctive emotional terrain. While we draw back from arguing for the possibility of a total regional geography of emotions, we would still claim that it is possible, desirable even, to argue for, and to understand more about, emotions that are socially placed. In the case of the Highlands, distinctive regional norms, expectations and cultures of emotional sociality make it possible to explore tentatively how emotional restraint, in particular, is something collectively produced and experienced (but while acknowledging important differences between different groups; see Parr *et al.*, 2004). In conclusion, though, we wish to recast our argument so far to highlight the centrality to our account of the processes generative of silence around emotions: processes that, to borrow from another vocabulary, appear to render emotion largely ineffable, 'unsayable', in the Scottish Highlands. In the model that has been our chief point of reference above, it is a mechanism of individual *and* collective repression that helps create 'silences' within psychological realms. Admittedly side-stepping complexities and contradictions of Freudian thought on this matter, we suspect that anything proving conflictual due to being *out* of kilter with everyday expectations, and hence a source of pain or mental anguish, is being consciously (or better pre-consciously) silenced. This means that repression is central to the making ineffable of this emotional 'stuff', rendering it basically unspeakable by the individual in their everyday settings, except possibly in encounters with experienced mental health workers.

Interestingly, a rather different intellectual tradition is now emerging in human geography to tackle the ineffability of emotions, but to do so through a more 'philosophical' lens and tied up with so-called 'non-representational theory' Thrift (1997, 1999, 2000a, 2000b: see also Smith, 2003). Claims about the fundamental *non*-correspondence between 'words and things' are central to this literature, and a more specific proposition is that the words of human language cannot adequately do the work of representing many interior mental and emotional states. In taking seriously the everyday performances through which humans 'think on [or with] their feet', coupled to their embodied immersion in a flow of conduct around and with other people, objects and worldly settings, the emphasis shifts from '*a priori* ideas' to 'the "earthiness" of our daily walks' as the placing of 'encounters for our tears and laughter' (Dewsbury, 2000, p. 493). Such immediate embodiment in the event-*full* everyday is seen as providing moments of emotional charge, release and possibility as a 'gap' or 'interval' anticipating the arrival of representation or, to put it another way, before any 'wording' of the sensation by the human subject either to themselves or to others. Within this gap *is* 'emotional tonality' (Varela, 1992: cited by Harrison, 2000, p. 509), and the further suggestion is that such emotional immediacy can disrupt 'habits' or 'styles' of existing thought and even prompt alternatives.

Reflecting more specifically on emotions in a non-representational mode, Harrison (2002, p. 3) elaborates as follows:

> Emotions, it seems relatively uncontroversial to claim, always threaten to overwhelm language beyond a communicative function. Emotions constantly threaten to disrupt and break-up our everyday use of language. To make use of [one] of Wittgenstein's phrases, emotions are perhaps exposed as language runs up against them just as language is exposed as they run through it ... However, this question by Wittgenstein indicates that this 'failure' of language, this inability of language to absorb emotions, may not be a failure of accuracy or representation, as if we simply lacked the 'right' words, as if the issue were simply to find the appropriate phrase or cliché then our problems would be solved. (Each of us knows already the terrifying banality of such clichés, and their unavoidability.)

Harrison speculates that the difficulty is 'not due to our lack of accurate concepts, but due to a constitutive resistance within emotions [to] words' and hence to 'meaning and representation' (p. 4). He also underlines that the response to this dilemma cannot simply be silence, but rather will usually entail a continual struggle to *try to* communicate in an emotional register, to utter speech striving to overcome the emotional chasm that resides – for instance – between the relative and the person on their deathbed or the relative and the person recently diagnosed with a mental health problem. In this non-representational guise, the study of emotional geographies must recognise that which cannot be said or, more positively, the diverse efforts – probably incorporating bodily gestures as well as tired clichés and embarrassed mumblings – that are witnessed being made in particular situations by particular people. At this point, there could well be traffic from Harrison's work to our own more empirical investigations of specific emotional terrains.

This being said, while a non-representational position traces this emotional ineffability to language's failings when confronted with the surplus of content seemingly endemic to an emotional register, we prefer to think in terms of different peoples in different places being more or less successful in their attempts to convey emotional states using ordinary language. While we agree that in a strict philosophical sense words can *never* be all that good at mirroring the fundamentally different 'stuff' that is emotion, we also take the famous Wittgensteinian dictum that 'the limits of our language are the limits of our world'[3] as a warrant for contemplating – in a quite empirical vein – how some worlds may, quite simply, be *more* limiting that others in what they permit those dwelling within them to say (see Smith, this volume). This means that it might be more pertinent to trace the unsayability of emotions in certain situations to the workings of local power relations, noting the greater or lesser influence of norms and expectations as bound up with which figures and institutions are effective locally, than to some distant philosophical impasse. In our Highland explorations, we have come across a place where various factors *do* indeed conspire to foster silence rather than talk around emotional issues, meaning that it is not inappropriate to speak of a local geography of emotional repression, but we also reckon that the story to be told elsewhere – on other emotional terrains – may well be very different and much more about what *is* and *can* be said rather than about what remains unsaid.

To offer a final summary, our chapter here reflects on the ordinary and widely-acknowledged social-cultural geographies through which, on some emotional terrains but maybe not all, the emotional register does become rather impoverished. Possibilities for emotional exchange are removed from such terrains, repressed by many of the people involved in accordance with assumed local norms and expectations, put away, bottled up by repressive practices that are themselves to some extent contingent on context. 'Not a display of emotions', the non-display of emotions, the non-speaking of emotions, the absence of emotional talk: these states are hence not to be explained solely by appeal to philosophy, or indeed by any singular appeal to a Freudian model of the unconscious. Instead, we suggest that this accounting must make at least some reference to what we are terming everyday emotional terrains, regions, places; *and* that in the process it must listen to what situated individuals within these places *do* manage consciously to 'say' about what they think is occurring. After all, our whole approach here, flawed as it may be, rests squarely on what our interviewees *felt* that they were able – and in practice *were* able – to represent to us about their own emotional worlds. This is not because we were great therapists or counsellors, although to some limited extent our interviewees may have responded to us in this way. Rather, it is because they could, at some level, if with some difficulty, often with a deep sadness, represent to us the realities, processes and consequences of emotional repression as it happens in a distinctive geographical setting. They may have revealed that strong emotions are rarely spoken about in their region, but they *could* nonetheless find the words, or at least some words, to speak about this socio-cultural version of repression, and this we would argue offers hope to those disadvantaged by the silences described previously, and tells us that geographies of emotions are both 'talk-able' and changeable.

Acknowledgements

This chapter draws on a research project funded by ESRC (R000238453). For details about the project, including findings and working papers, see the project website at http://www.geog.gla.ac.uk/Projects/WebSite/main.htm. Thanks as well to the editors for their advice and encouragement, and to Paul Harrison for allowing us a sight of his 2002 typescript.

Notes

1 The Hebrides is the collective name for groups of islands off the west coast of mainland Scotland and form an integral part of the Scottish Highlands, sometimes called the Highlands and Islands.
2 Repression has to be understood in relation to the 'pleasure principle': 'an innate tendency to want to maximise pleasure and minimise pain' (Badcock, 1988, p. 16). Moreover, '[i]f we take the dynamic foundation of the topographical view of the conscious seriously we must conclude that mental topography – the real, active distinction between conscious, pre-conscious and unconscious – is maintained by the force of repression and that this, in turn, exists to attempt to safeguard consciousness from conflict, anxiety and contradiction' (p. 16).
3 To paraphrase the original: 'The limits of my language mean the limits of my world' (a phrase repeatedly revisited by the likes of Olsson (eg. 1980) and also now used by the likes of Harrison (2000, 2002)).

References

Agnew J (1996) 'Liminal Travellers: Hebridians at Home and Away' *Scotlands*, 3, 1, pp. 32-41.
Agyeman J and Spooner R (1997) 'Ethnicity and the rural environment' in Cloke P and Little J (eds) *Contested Countryside Cultures: Otherness, Marginality and Rurality.* (Routledge, London), pp. 197-217.

Badcock C (1998) *Essential Freud.* (Basil Blackwell, Oxford).

Bell D and Valentine G (1995) *Mapping Desire: Geographies of Sexualities.* (Routledge, London).

Brown GW and Harris T (1978) *Social Origins of Depression: a Study of Psychiatric Disorder in Women.* Tavistock Publications, London.

Brown GW and Prudo R (1981) Psychiatric disorder in a rural and an urban population: 1. Aetiology of depression. *Psychological Medicine* 11, 581-599.

Brown GW, Davidson S, Harris T, Maclean U, Pollock S and Prudo R (1977) Psychiatric disorder in London and North Uist. *Social Science and Medicine* 11, 367-377.

Callard F (2003) 'The taming of psychoanalysis in Geography' *Social and Cultural Geography*, 4, pp. 283-293.

Dewsbury J-D (2000) 'Performativity and the event: enacting a philosophy of difference' *Environment and Planning D: Society and Space* 18 473-496.

Erwin E (2002) *The Freud Encyclopedia: Theory, Therapy and Culture.* (London: Routledge).

Free Church of Scotland (1998) 'Religion and mental illness' *The Monthly Record of the Free Church of Scotland* January, pp. 3-5.

Fuller J, Edwards J, Procter N and Moss J (2000) 'How definition of mental health problems can influence help seeking in rural and remote communities' *Australian Journal of Rural Health* 8 148-153.

Harrison P (2000) 'Making sense: embodiment and the sensibilities of the everyday' *Environment and Planning D: Society and Space* 18 497-517.

Harrison P (2002) '"How shall I say it ...?" emotions, exposure and compassion' Paper given at *Emotional Geographies Conference*, Lancaster 2002, typescript provided by author.

Hunter J (1995) *On the Other Side of Sorrow: Nature and People in the Scottish Highlands.* (Mainstream, Edinburgh).

Macdonald S (1994) 'Whisky, women and the Scottish drink problem: a view from the Highlands' in MacDonald M (ed.) *Gender, Drink and Drugs.* (Oxford: Berg), pp. 125-143.

Macritchie STM (1994) Celtic culture, Calvinism, social and mental health on the Island of Lewis. *Journal of Religion and Health* 33(3), 269-278.

McDonald F (2000) 'Producing space in Presbyterian Scotland: Highland worship in theory and practice' *Arkleton Research Papers.* No.2. University of Aberdeen.

Olsson G (1980) *Eggs in Bird/Birds in Egg.* (Pion, London).

Parman S (1990) *Scottish Crofters: a Historical Ethnography of a Celtic Village.* (Holt, Rinehart and Winston, Fort Worth).

Parr H and Philo C (2003a) 'Introducing Psychoanalytic geographies' *Social and Cultural Geography*, 4, pp. 283-293.

Parr H and Philo C (2003b) 'Rural mental health and social geographies of caring' *Social and Cultural Geography* 4, pp. 471-488.

Parr H, Philo C and Burns N (2004) 'Social geographies of rural mental health: experiencing inclusion and exclusion' *Transactions of the Institute of British Geographers* 29, 401-419.

Philo C, Parr H and Burns N (2003) 'Rural madness: a geographical critique of the rural mental health literature' *Journal of Rural Studies* 19 259-281.

Pile S (1996) *The Body and the City: Psychoanalysis, Space and Subjectivity.* (Routledge, London).

Prudo R, Harris T and Brown GW (1984) Psychiatric Disorder in a rural and an urban population: 3. Social integration and the morphology of affective disorder. *Psychological Medicine* 14, pp. 327-345.

Prudo R, Brown GW, Harris T and Dowland J (1981) Psychiatric disorder in a rural and an urban population: 2. Sensitivity to loss. *Psychological Medicine* 11, 601-616.

Rycroft C (1995) *A Critical Dictionary of Psychoanalysis.* (Penguin, London).

Smith RG (2003) 'Baudrillard's nonrepresentational theory: burns the signs and journey without maps' *Environment and Planning D: Society and Space*, 21, pp. 67-84.

Thien D (2005) *Intimate Distances: Geographies of Gender and Emotion in Shetland.* (Unpublished PhD thesis, University of Edinburgh).

Thrift N (1997) 'The still point: expressive embodiment and dance' in Pile S and Keith M (eds), *Geographies of Resistance.* (Routledge, London), pp. 124-51.

Thrift N (1999) 'Steps to an Ecology of Place' in Massey D, Allen J and Sarre P (eds) *Human Geography Today.* (Polity Press, Cambridge), pp. 295-322.

Thrift N (2000a) 'Non-representational theory' in Johnston R, Gregory D, Pratt G and Watts M (eds) *Dictionary of Human Geography: 4th Edition.* (Blackwell, Oxford), pp. 556.

Thrift N (2000a) 'Afterwords' *Environment and Planning D: Society and Space*, 18, pp. 213-256.

Chapter 8

Freedom, Space and Perspective: Moving Encounters with Other Ecologies

David Conradson

Introduction

For many people in the western world, spending time in scenic natural surroundings is a valued counterpoint to the demands of work and home life. Whether undertaking a particular activity – such as walking, gardening or cycling – or simply 'being present' in a less directed fashion, these environmental encounters are in part appreciated for their capacity to move us to think and feel differently (see Milligan *et al.*, this volume). In coming close to other ecologies and rhythms of life, we may obtain distance from everyday routines, whilst perhaps also experiencing renewed energy and finding different perspectives upon our circumstances. These emotional gains are one reason why such environmental encounters are both prized and the focus of significant commodification. They are feelings which arguably help sustain particular traditions of self-landscape engagement.

In this chapter I explore the emotional dimensions of such encounters with particular reference to a landscape in Dorset, southern England. The immediate focus is a respite care centre, an agency that seeks to facilitate rest and wellbeing for its guests, the majority of whom experience some form of physical impairment. In addition to the residential accommodation and care provided, staff also place significant emphasis upon facilitating access to the wider natural environment in which the organisation is located. This landscape, which consists of 350 acres of wooded heathland, bounded on one side by a large natural harbour, is seen as integral to the agency's efforts to care for those who visit. As I elaborate later, engagement with its component parts – the trees, birds, deer and other wildlife – is thus encouraged and enabled in a series of ways.

The discussion that follows is guided by two main questions. Firstly, what might it mean to conceptualise the engagement between the self and landscape as a relational encounter? The response I develop here draws on ideas from both psychotherapy and recent work in human geography on ecologies of place. This theoretical juxtaposition reflects a broader and ongoing interest in the utility of relational and psychotherapeutic approaches for explicating the felt dimensions of encounters with place (Conradson 2003a, 2003b). Moving to the case-study, the second question is then: what emotional dynamics emerge for individuals who find themselves imbricated within the particular landscape of Holton Lee? Or, put another way, how does dwelling within this setting appear to shape a person's subjectivity, both during a visit and after departing? With reference to the spoken narratives of a selection of guests, here I highlight some of the ways in which the landscape's physical extensivity appeared to draw out – or at least facilitate – feelings of internal spaciousness that were often described using terms such as 'freedom'. Guests approached and experienced the landscape in quite different ways, however, and consideration is given to the issue of individual receptivity in this regard. A brief conclusion then draws together the main points.

Landscape, Encounter and Feeling

Within Western culture, there is a long tradition of regarding certain natural landscapes as potentially therapeutic settings (Gesler, 2003). Some of the areas valued in this way have included forested downlands, mountain environments, coastal zones and geothermal regions (Dobbs, 1997; Palka, 1999; Kaplan and Talbot, 1983). In each case, 'nature' is of course neither an unproblematic category nor necessarily benign. And there are other important modalities of landscape engagement too, including those oriented around curiosity, conquest (e.g. exploration and mountaineering) and aesthetic cultivation (e.g. landscape gardening). But the popularity of practices in which people 'retreat to nature' in search of refreshment, perspective and even personal renewal continues. Such activity can be observed in a wide range of settings today, from urban parks and green spaces all the way through to lakes and alpine landscapes. Direct physical engagement with the constituent elements of these environments is often experienced as calming, eliciting comments on the 'peacefulness' or 'restfulness' of such places (Hartig *et al.*, 1991; Williams, 1999; Gerlach-Spriggs *et al.*, 1998).

One way of understanding these emotional experiences is as *affective* outcomes of relational encounters between the self and a landscape. With origins that span psychiatry, cognitive science and cultural theory, the notion of 'affect' is used here – as it has been in other recent human geographic writings – to refer to shifts in the energetic capabilities of a body, in a manner which transcends but nevertheless remains attentive to shifting contours of feeling (McCormack, 2003; Thrift, 2004; Harrison, 2004). As Deleuze and Guattari (1987: xvi) explain in the introduction to *A Thousand Plateaus*:

> Neither word [affect/affection] denotes a personal feeling ... [Affect] is a pre-personal intensity corresponding to the passage from one experiential state of the body to another and implying an augmentation or diminution in that body's capacity to act.

In writing of pre-personal intensity, the formulation of affect here is akin to a field, or line of force, that exists *prior* to the individual who encounters it. When an individual states they feel happy or energised, for example, Deleuze and Guattari invite us to consider this feeling as reflective of an individual's encounter with a broader, supra-individual or transhuman affective field. As Thrift (2004: 60) elaborates, to think about affect is to:

> ... set aside approaches that tend to work with a notion of individualised emotions (such as are often found in certain forms of empirical sociology and psychology) and stick with approaches that work with a notion of broad tendencies and lines of force: emotion as motion, both literally and figurally.

In comparison to individualised formulations of emotion, affect is thus more attentive to both the embodied and intersubjective dimensions of human feeling.

Before exploring the affective dimensions of the self-landscape encounter, it is useful to consider the relational nature of both the self and landscape. With regard to the self, psychotherapy offers valuable insights, for it is a body of thought that pays close attention to the psychosocial dynamics of subjectivity and emotion. As in psychoanalytic thought, psychotherapeutic writings are characterised by a range of different analytical and theoretical stances (Clarkson, 1995; Feltham, 1999; Nelson-Jones, 2001). Within the three broad traditions of contemporary psychotherapy – the psychodynamic, cognitive-behavioural and humanistic-existential – there is, nevertheless, a common interest in understanding the human self as a relational entity (Hazler and Barwick, 2001; Bondi, 2003). In contrast to tightly bounded or autonomous formulations of personhood, the self is here understood as emerging within and through its relations to other people and events. At the broadest level, the practice of psychotherapy recognises that the nature of these interrelations – their content and experiential dynamics – has the potential to shape the person in both positive and negative ways. Furthermore, psychotherapists contend that these relations can be worked upon, with a view to enabling the self to unfold in different, more positive ways. The relations of concern may be located in the present, in the form of ongoing friendships for instance, but equally they may involve connections to people and events in other times and places. The relational self is thus a distributed entity, in that the faculties of memory and imagination permit the maintenance of connections across a range of times and spaces. In this sense, Hägerstrand's (1982) famous time-space diagrams, which depicted the self moving through linear time, represent the physical body accurately but fail to map the rather more dispersed spatiality of the internal self (Bondi with Fewell, 2003).

This psychotherapeutic perspective allows us to begin to see how our emotional lives are in part shaped by wider relational contexts. For the self is understood as a malleable and somewhat porous entity, able to be affected by others, in part by internalising its relations with them in various ways. In additional to direct, material impacts, interpretation is of course also important here, in that our perception of an event or relation is often as significant in determining its impact upon us as any catalogue of its raw characteristics. What is comfortable or acceptable for one person may thus be distasteful or even distressing for another. To acknowledge the importance of relational context for human emotional dynamics in this way does not require subscription to a thoroughgoing social constructionism, however, where the person becomes merely an effect of external discourses and practices. Whilst acknowledging the influence of relationality, we can instead contend that the self exhibits a degree of centring and consistency within the relations that shape it. Whether this core is best narrated in biological, psychological or spiritual terms – or some mix of the three – is a matter of debate, but the principal point is that the self then becomes a *somewhat* centred entity that emerges through a reflexive and relational interplay with other people and events.

Attending to psychosocial relations clearly takes us some way towards understanding the emotional dynamics of the self (Clarkson, 1995; Rogers, 1957, 1980). But if we wish to examine the felt dimensions of a person's encounter with a landscape, then a conception of relationality that extends beyond social interaction alone is needed. For places are of course comprised not only of people, but also of a plethora of non-human life forms and inanimate objects.

Nigel Thrift's (1999) recent work on the *ecology of place* is a useful point of reference in this regard. Drawing inspiration in part from actor-network theory, Thrift employs the term ecology to signal an interest in the broad spectrum of entities that comprise a place, and the interactions between them. While actor-network theory has usefully directed our attention to the significance of other life-forms and objects in our everyday worlds, its ambivalence about the concept of place is also a limitation. As Thrift (1999, p. 313) notes:

> ... Latour and other actor network theorists often fail to see the importance of place: their vision of a radically symmetrical network of networks, consisting of different aspects like humans, animals and things, and mobiles like writing, print, paper and money constantly combining and recombining is an important corrective to simple humanisms and to simple notions of connectedness, but it also means that actor network theory cannot speak of certain things. In particular, Latour and other actor network theorists often fail to see the importance of place because they are reluctant to ascribe different competencies to different aspects of a network or to understand the role of common ground in how networks echo back and forth, often unwittingly.

Thrift therefore works with a more general conception of relational materialism, eschewing these limitations of the network metaphor. In promoting awareness of non-human others, along with inanimate and technological objects, an ecological conception of place shifts the frame of reference beyond humans alone. It asks us to remain attentive to social interactions (between people), but it also invites consideration of the various forms of engagement between people and things (e.g. with books, cars, telephones), and between people and other biological entities (e.g. food, trees, microbes, pets). Extending the earlier point about the porosity of the self, an ecological perspective highlights the significant material exchanges between people and their environments, whether through ingestion of food, respiration, excretion or more complex engagements with techno-material objects. For as Thrift (1999, p. 312) writes, things are 'folded into the human world in all manner of active and inseparable ways, and most especially in the innumerable interactions between things and bodies'.

While these interactions are in one sense material, for humans they may also be associated with particular affective and emotional outcomes. In reflection of a broadly post-humanist stance, Thrift (1999) considers emotion less as the bounded property of an individual actor and more in terms of intersubjective fields of affective intensity that emerge between individuals (see also Thrift, 1996, 1999, 2004). This is not about denying that people feel – for of course they do – but instead an argument that feeling is not only or entirely personal (or individual). From an affectual point of view, feeling is something that may emerge *between* bodies of various kinds, whether human or otherwise.

While an ecology of place thus acknowledges what might be described as individualised formulations of emotion, it also goes further. In particular, while in an actor network perspective 'the way in which things are mixed in can seem prosaic', an ecology of place seeks to be attentive to those complex dynamics of feeling that sometimes lie 'at the edge of semantic availability' (Thrift 1999, p. 313). Beyond individual happiness or sadness, this is about witnessing the complex affections, ambivalences and antipathies that at times pass through places. Although emerging from a different, perhaps even incommensurate theoretical tradition, there are arguably resonances here with the concerns of humanistic geographers such as Tuan (1976, 1977), Relph (1981) and Seamon (1982, 1984; Seamon and Maugerauer, 1989) who, with their phenomenological and other social theoretic insights, sought to probe human feelings about and attachments to place. While its period of primary influence has now passed, elements of this humanistic geographic impulse have undoubtedly been incorporated into the new cultural geography (see, for example, Philo 1991; Duncan and Ley 1993; Cook *et al.* 2000), and in various ways thus live on.

To summarise thus far, the discussion of self-landscape encounters undertaken here seeks to integrate a psychotherapeutically derived, relational conception of self with an ecological conception of place. The self is understood as emerging in part through its relations with other people and events, whether these are present in the 'here and now' or located in the 'there and then'. These influential others need not just be people or events, however, but may also include the various non-human life forms and inanimate things that comprise a place or landscape. As an example of this style of work, we might consider Cloke and Jones' (2001, 2004) recent discussions of the complex material and emotional associations that develop between humans and trees in particular places. As individuals become imbricated within particular ecologies of place, so their emotions – whether happiness, sadness, elation, gloom, relief or anger – arise in part from embodied physiological and psychosocial responses to the constituent elements of those places. This is not a matter of environmental determinism (see Smith, this volume), for the relations are typically characterised by interplay rather than unidirectional forcing, and in any case people exhibit a degree of consistency across different settings. Nonetheless, places do have the capacity to shape our feelings.

To move between places is thus to transition between different ecologies of people, bodies and things. Such movement is of course part of the fabric and rhythm of everyday life: people travel from domestic settings to work, for instance, perhaps taking in other locations during the day, before returning to a home-place in the evening. While much everyday movement is undertaken as a means to other ends – to complete a job, to collect a child, to visit the doctor and so forth – there are also instances where individuals move self-consciously in an effort to work upon their emotions. They may go to the park for a walk, or perhaps relocate to a favourite place for a period of time, so as to calm down and obtain some perspective after a stressful day in the office (Korpela and Hartig, 1996). Such movement arguably involves a degree of recognition – whether conscious or otherwise – of the effect that one's current ecology of place is having in emotional terms, as well as the potential effects that changing this ecology might bring. In the following section, these ideas are used to explore the emotional experiences of individuals at a particular landscape in southern England.

Moving Encounters: Negotiating an English Landscape

Set in an extensive area of wooded heathland in rural Dorset, Holton Lee is a respite care centre with a primary focus on individuals who experience various forms of physical impairment. In 2004, around 25 staff and 40 volunteers were involved in the daily operation and upkeep of the organisation. Their core focus, as the mission statement explains, is 'empowering and resourcing people, particularly carers and disabled people, through creativity, environmental awareness, personal growth and spirituality'. The residential accommodation and care facility, known as the Barn, is central to this work. At present it is able to host up to twelve individuals at a time, with personalised assistance provided by a well-qualified group of care staff. Guests typically stay for a week per visit, with some individuals visiting up to six times a year, depending on the nature of their respite care needs and the levels of financial assistance they can obtain.

The other major element of Holton Lee's efforts to facilitate personal wellbeing consists of enabling access to the 350 acres of wooded heathland it is set within. This land had previously belonged to a local family as part of working farm, but in 1992 it was gifted to a charitable trust with a vision of developing a centre for holistic rest and renewal. To help individuals negotiate the landscape, a series of gravelled paths were laid across the site. These lead from the accommodation to a series of bird hides – from where individuals can observe some of the rich and varied local wildlife – as well as enabling more general exploration of the heathland. Whilst suitable for wheelchairs, in view of the distances involved and so that guests can travel cross-country, there are also a series of small electric vehicles with the trade-name of *Tramper* that allow independent travel.

The discussion of Holton Lee here is based on a series of research visits between September 2002 and February 2004, undertaken as part of a broader consideration of places of retreat in contemporary Britain. The primary aim was to develop an understanding of guests' experiences while they visited Holton Lee for respite care. Following an initial meeting with the Director and a member of the Board of Trustees, the approach adopted for making contact with guests involved the Accommodation Manager distributing a short letter to them upon arrival. This letter summarised the project and asked if they would be willing to participate in a semi-structured interview about their experience of visiting. Conversations were arranged and conducted with 22 guests in this way, and these were complemented by interviews with six staff that explored the organisation's history, values and service practices. The final sample of guests was comprised of an equal number of men and women, with an average age of around 45.[1] Over 80 per cent of respondents made at least occasional use of a wheelchair for mobility purposes, in association with conditions such as multiple sclerosis or as a result of having experienced strokes, spinal injuries or other musculo-skeletal problems. All interviews were taped and then transcribed for subsequent thematic analysis.[2]

Experiencing Other Ecologies

Many guests understood their time at Holton Lee as being about temporary imbrication within an ecology of place that differed from that of their home environments. For a start, they saw their visits as an opportunity to obtain distance from everyday routines and domestic demands. There was a sense of appreciation at having various practical jobs done for them, for instance, such as cooking, cleaning and washing. In comparison to domestic settings where there was often some reliance on care assistants for such tasks, Holton Lee was often experienced as somewhere with greater scope for flexible and personalised attention. Its relatively small bed-numbers and ethos of hotel standard service were significant in fostering a professional yet restful environment. As Cameron,[3] a middle-aged guest, explained:

> It helps you to relax. At home I can't relax. This is just a break away from the mundane, you know, sort of humdrum. You can forget about the rest of, you know, the rest of whatever. And it is therapeutic, yeah. You come here and relax.

Being away from home was of course about more than relief from domestic tasks. Crucially, Holton Lee also afforded guests some distance from demands and expectations that issued from beyond their home environments. Elsie, who was extensively involved in voluntary organisations in her local area, thus noted that:

> I look forward to it as a time of relaxation, of being just me. I look forward to it as a place where the pressure is lifted. Here I am away from my phone. I don't have to look at my diary. I can just be me. I can relax and enjoy just doing what I want to do. There is no pressure from anybody else. You take the lid off the pressure cooker, which then gives me the strength to go back and carry on.

Her comment about finding the 'strength to go back and carry on' is illustrative of the way many guests looked to Holton Lee to provide refreshment. This is not to suggest that Holton Lee was without its own set of social expectations and norms, but rather that those it did have appeared to allow some guests to unwind in a different fashion than was the case in their home settings.

As noted earlier, the heathland at Holton Lee occupied a significant place in the organisational strategy for facilitating health and wellbeing. Many interviewees made reference to the spaciousness of the site in this regard, both in landscape terms but also with respect to the 'internal' room they felt it afforded them.

Referring to the electric vehicles available for independent exploration of the site, Cameron explained that during his visits he would spend:

> All the time I can get on the Tramper [electric vehicle]. I can go through the woods or get onto the Heath. So, you know, it just expands your horizons. It's the one place where you can get a sense of freedom. ... With the off road vehicles they've got, the buggies and that, it just means that I've got a certain degree of freedom that I don't have in the rest of my life.

Here we see an experience of physical extensivity crossing over to a particular sense of emotional expansiveness. According to the staff who worked out-of-doors, 'freedom' was in fact a term used relatively often by guests when describing their feelings about time on the heath. In addition to the dimensions of the physical landscape, contact with wildlife was particularly valued in this regard. It was common to see deer on the site for instance; having become accustomed to human presence, they would congregate quite regularly in the fields outside the Barn at dusk or dawn. The birds, badgers, foxes and various aquatic animals were also noted. Of course, these encounters were never about interaction with 'first nature'; the site had long been managed for both agricultural and recreational purposes, and it was not in any sense a wilderness. Nevertheless many interviewees found the extensivity, relative quiet and rich variety of flora and fauna of the Holton Lee landscape to be instrumental in the emergence of a different set of feelings than those they typically experienced in their home settings. Sometimes this seemed to reflect the landscape's resonance with their notions of rural idyll, but pleasant memories of similar coastal heathland settings and the documented mood-enhancing effects of such environments were also important (Hartig *et al.*, 2003).

Inhabiting Open Space

Guests inhabited the spaciousness of the site in different ways. Some used it as an opportunity to slow down, re-centre and reflect upon their lives. This was the experience of Elsie, who felt that at Holton Lee, 'you've got space and time to enjoy nature. ... You can take time out without feeling guilty, so that you can relax and really look at your life, look at your self, look at what you're doing'. For Paul, the effect was somewhat different; he found the landscape drew him 'out of himself', away from internal ruminations and medical scrutiny regarding his physical health problems:

> I think it makes you forget about yourself for a start. You can get so cramped in, on housing estates and towns and things like that. But to come here, you can see for miles. And birds are flying around, rabbits, foxes and deer. ... And you know the atmosphere outside is just peaceful. It's always just peaceful. You sort of come up the road here, that's when we start relaxing as soon as we come up the road.

Like several other guests, Paul was relatively active in exploring the site, whilst also taking time to read and participate in some of the creative classes offered (e.g. painting, sculpture). His was a relatively outwardly focused form of re-energisation.

In relational terms, these affective outcomes – shifts in bodily energy and feeling – emerged in part through the opportunity guests' had to attain some distance from the immediate demands and expectations of their home environments. Equally significant, however, were the green space, physical extensivity, relative quietness and opportunity to come close to a range of animals. This is not to imply that all guests were equally involved in exploring the site – a small number spent the majority of their visit within the residential accommodation, resting, socialising with other guests and watching television for instance. For those who did explore the landscape, however, they generally did so because of the enjoyment and relaxation it afforded. These kinds of therapeutic effects have been well documented by environmental psychologists, principally as part of the so-called restorative settings literature (Hartig and Staats, 2003; Hartig *et al.*, 2003; Herzog *et al.*, 2002). Similar observations, generated in a less experimental manner, also exist in the literature on therapeutic landscapes (see Williams, 1999 for an overview) and within health and landscape architecture (Gerlach-Spriggs *et al.*, 1998; Marcus and Barnes, 1999; Tyson, 1998).

Reflecting its ability to function for them as a restful and restorative place, most guests sought to schedule as many of their respite care visits at Holton Lee as possible. At a cost of between £300 and £500 per week in 2003 (dependent on the level of personal assistance required), however, realising these aspirations typically required financial assistance from government social security. Holton Lee's costs were in the upper half of the expense range for voluntary and statutory respite care facilities in the Dorset area, and while guests were typically convinced of its merits over more homogenised or larger-scale services, funders were often eager to minimise the bottom line expense. Each party, as Cameron explained, evaluated the costs and benefits differently:

> I think that Social Services treat [Holton Lee] as an expensive resource. But for disabled people who use it, um, the benefits outweigh the expense. They do have a new day centre now in Wareham which they tried to get us all to go to and, um, it's just totally unsuitable. *What's unsuitable about it?* Ah, location, staff, attitude. We're all, we were invited to the new day centre to have a look round. And really I was there five minutes. It's opposite the vets and alongside the railway line, just outside of Wareham on an industrial estate. I asked, you know, would I be able to do bird-watching here and things? And they said, well it depends on the birds. And I said, well what about photography? They said well if I've got a camera. Plus the other unsuitable thing about it is that you don't get a meal. You've either got to bring sandwiches or bring something to cook in their kitchen. They don't supply anything else, like Holton Lee do. And this is a Social Services, Dorset County Council run centre.

Although Cameron emphasised the activities unavailable to him at the day care setting, his broader interview narrative suggested his concerns were also about what their absence would mean in terms of *feelings*. In his experience, access to the Holton Lee environment was arguably about access to a particular emotional domain, an affective realm that for practical reasons was less easy to enter elsewhere. It was an environment which afforded him a certain sense of freedom and internal respite. From his perspective, the emotional dynamics of visiting Holton Lee were arguably as important as the material care and support it offered.

Given that many guests shared this preference for Holton Lee over the local statutory care services, the thought of no longer having access was not at all attractive. When asked how she would feel if no longer able to visit, Caitlin responded as follows:

> Oh, no, I'd hate it. ... I'd be losing friendship. Umm. The peace, getting away from it all, which is lovely. You even get away from my husband's snoring!

Other interviewees spoke in less light-hearted terms, noting the difficult feelings that periodically surfaced as a result of limited physical mobility, dependence upon others for aspects of personal care, and at times experiencing a degree of social isolation. Holton Lee was then a place that they used to negotiate these feelings. As Cameron explained:

> I get cabin fever [at home]. It's not quite that bad, but it's fairly difficult. I mean this [Holton Lee], literally, this is the safety valve. This is what keeps me sane. ... I don't even want to think about not coming to Holton Lee. It is, it is that safety valve. It really is.

There was a clear sense here, reiterated by other guests, that respite care at Holton Lee played an important role in managing the emotional dynamics associated with being partially mobile and in some way dependent upon care. Several interviewees spoke of looking forward to visits in advance and of being disappointed, struggling even, if for some reason they were unable to attend after all.

Temporalities of Affect: Attenuation and Maintenance

For most guests, the sense of equilibrium attained through visiting Holton Lee would stay with them for a little while after departure, but the feelings would eventually begin to attenuate. Between visits, many individuals thus sought to keep the Holton Lee landscape 'alive' in various ways. Couples would share reminiscences, friends would make phonecalls, whilst others would recall images and feelings from the landscape. As Elsie put it:

> If you're feeling a bit down you can just shut your eyes and picture the heath land, picture looking over Poole Harbour, you can picture the sea coming in and taking your troubles out with the sand. ... The trees, the heathland, the wind. ... The wonderful deer dignified just walking through the trees. You can take that with you and if life gets a bit hectic and frantic you can just say, hang on a minute, let's just think about that for a minute. And you can then let everything down.

In a series of ways then, spending time at Holton Lee might be said to have moved the respondents to think and feel differently about themselves and their life opportunities. Some individuals were able to internalise these positive feelings, to in a sense fold them into themselves, thereby extending the temporal duration of the landscape's affective influence upon them.

In accounting for different experiences of the Holton Lee landscape, there is perhaps a point to be made about receptivity. Unsurprisingly, there was a significant spectrum of attitudes amongst guests in terms of their enthusiasm about being at Holton Lee. On a few occasions, individuals were temporarily accommodated as a result of Social Services shortage of appropriate alternative accommodation for instance; their residence was not voluntary. Amongst those who had intentionally chosen to spend time in the Barn, there was still typically a process of initial adjustment to the routines and new environment. Sometimes Holton Lee would 'work its magic', as one guest put it, irrespective of an individual's degree of openness or enthusiasm, in a sense dissolving defensive or resistant internal postures over time, perhaps enabling them to find a freer exchange of feeling and energy within themselves and with others. In other cases, however, a person might be unable or less willing to engage with the environment in this way. As Paul explained:

> Some people are so, well if you're strung out when you first come to this place, well it brings you back down to earth I think. Unless you're *too* strung out to appreciate it. There's that to it as well. Sometimes you see people that, they don't seem to appreciate it. And they, um, spend all day watching the TV. You can do that at home.

In relational terms, we might say that despite their physical imbrication within the Holton Lee landscape – and thus its proximity – these individuals engaged less fully, or perhaps somewhat reluctantly, with their immediate ecology. In these circumstances, it is possible that the emotional contours of their visit may have been shaped less by their surroundings, and a little more by the relations carried within them, internalised from other places. The Holton Lee landscape they encountered may thus have felt quite different to the place reported by some of the other guests in this chapter.

Conclusions

In considering the experience of guests staying at Holton Lee, this chapter has sought to adopt a relational approach to the self-landscape encounter. Central to the argument was an ecological conception of place, a perspective which takes place to be a rich constellation of human, non-human and material entities, coinciding and interacting with each other (Thrift, 1999). The self is imbricated within and shaped by its relations with these entities; as an extension of psychotherapeutic arguments would suggest, these people, non-humans and things give form to both its present and potential dimensions (Clarkson, 1985; Hazler and Barwick, 2001; Bondi, 2003). If we accept that places are comprised of relatively consistent assemblages of these entities, then it becomes possible to see how moving between settings has the potential to facilitate new emotional dynamics for those involved.

Across the interviewees, the emotional experience of spending a week or two in the Holton Lee environment was typically positive. This is not to imply uniformity for, as noted, guests inhabited the landscape's extensivity in different ways, ranging from being relatively reflective through to adopting an energetic outward focus accompanied by significant physical activity. In general, however, Holton Lee was experienced in terms of relative freedom and renewed perspective. As guests interacted with the landscape through practices such as exploring and bird-watching, a sense of calm often emerged. These feelings were something individuals often then sought to preserve and keep alive when they returned to their domestic environments.

When the self is understood as a complex relational entity, capable of making and sustaining connections with other bodies across a range of times and places, then it becomes possible to appreciate how context exercises some influence upon human feeling. Such an approach certainly aids our understanding of the felt dimensions of self-landscape encounters, whilst also helping us to account for variation in individual experience and interpretation of particular landscape elements. It also highlights the contemporary significance of mobility and relocation for the reflexive management of one's emotions.

Acknowledgements

I am grateful to the editors for their helpful comments on an earlier version of this chapter, and to Carl Griffin, Alan Latham and Derek McCormack for their thoughts on the arguments developed here. Elements of the work have also benefited from the responses of seminar audiences at the Universities of Dundee, Edinburgh, Washington and New South Wales.

Notes

1　An approximate figure, as not all interviewees disclosed their age in precise terms.
2　As with most intensive qualitative research, the sample of guests here was not intended to be representative. Those who volunteered to participate in the study were essentially self-selected on the basis of being both willing to speak with a (relative) stranger and well enough to do so. Given the usual social inhibitions around expressing negative opinions to strangers, it is likely that individuals with less than positive experiences of Holton Lee were thus under-represented in the sample. For the purposes of this study, this was non-critical however, as the primary interest was in the relational dynamics which underlay the emotional dimensions of their time at Holton Lee.
3　All names of guests used in the text are pseudonyms.

References

Bondi, Liz with Judith Fewell. (2003) 'Unlocking the cage door': the spatiality of counselling, *Social and Cultural Geography* 4, 527-547.

Bondi, Liz (2003) 'A situated practice for (re)situating selves: trainee counsellors and the promise of counselling' *Environment and Planning A* 35: 853-870.

Clarkson, P. (1995) *The Therapeutic Relationship in Psychoanalysis, Counselling Psychology and Psychotherapy*. Whurr Publishers, London.

Cloke, P. and Jones, O. (2001) 'Dwelling, place and landscape: an orchard in Somerset', *Environment and Planning A* 33: 649-666.

Cloke, P. and Jones, O. (2004) 'Grounding Ethical Mindfulness for/in Nature: Trees in Their Places', *Ethics, Place and Environment* 6(3): 195-214.

Conradson, D. (2003a) 'Spaces of Care in the City: The Place of Drop-In Centres', *Social and Cultural Geography* 4(4): 507-525.

Conradson, D. (2003b) 'Doing organisational space: practices of voluntary welfare in the city', *Environment and Planning A 35*: 1975-1992.

Cook, I., Crouch, D., Naylor, S. and Ryan, J.R. (eds) (2000) *Cultural Turns/Geographical Turns*. Harlow, Pearson.

Deleuze, G. and Guattari, F. (1987) *A Thousand Plateaus*. London: Athlone Press.

Dobbs, G.R. (1997) Interpreting the Navajo sacred geography as a landscape of healing. Pennsylvania Geographer 35(2), 136-150.

Duncan, J. and Ley, D. (1993) *Place/culture/representation*. London, Routledge.

Feltham, C. (ed.) (1999) *Understanding the Counselling Relationship*. Sage, London.

Gerlach-Spriggs, N., Kaufman, R.M. and Warner, S.B. (1998) *Restorative Gardens: The Healing Landscape*. Yale University Press, Yale.

Gesler, W. (2003) *Healing Places*. Rowman and Littleford. Lanham MD.

Hägerstrand, T. (1982) 'Diorama, path and project', *Tijdschrift voor Economische en Sociale Geografie* 73: 323-39.

Harrison, P. (2004) 'What is affect?' Online submission to Critical Geography Forum, 29th June. Accessible from www.jiscmail.ac.uk.

Hartig, T., Mang, M. and Evans, G.W. (1991) 'Restorative effects of natural environment experiences', *Environment and Behaviour* 23: 3-26.

Hartig, T. and Staats, H. (2003) 'Guest Editors Introduction: Restorative Environments', *Journal of Environmental Psychology* 23: 103-107.

Hartig, T., Evans, G.W., Jamner, L.D., Davis, D.S. and Gärling, T. (2003) 'Tracking restoration in natural and urban field settings', *Journal of Environmental Psychology* 23: 109-123.

Hazler, R. and Barwick, N. (2001) *The Therapeutic Environment: Core Conditions for Facilitating Therapy*. Buckingham: Open University Press.

Herzog, T.R., Chen H.C. and Primeau, J.S. (2002) 'Perception of the Restorative Potential of Natural and Other Settings', *Journal of Environmental Psychology* 22: 295-306.

Kaplan, S. and Talbot, J.F. (1983) 'Psychological benefits of a wilderness experience', in Altman, I. and Wohlwill, J.F. (eds) *Human Behaviour and Environment: Advances in Theory and Research, volume 6: Behaviour and the Natural Environment*. New York: Plenum Press, pp. 163-203.

Korpela, K. and Hartig, T. (1996) 'Restorative Qualities of Favourite Places', *Journal of Environmental Psychology* 16: 221-233.

Marcus, C.C. and Barnes, M. (1999) (eds) *Healing gardens: therapeutic benefits and design recommendations*. Wiley, New York.

McCormack, D. (2003) 'An event of geographical ethics in spaces of affect', *Transactions of the Institute of British Geographers* 28(4): 488-507.

Nelson-Jones, R. (2001) *Theory and Practice of Counselling and Therapy*. London: Continuum.

Palka, E. (1999) Accessible wilderness as a therapeutic landscape: experiencing the nature of Denali National Park, Alaska. In: Williams, A. (Ed.) *Therapeutic Landscapes: The Dynamic between Place and Wellness*. University Press of America, Lanham, pp. 29-52.

Philo, C. (ed.) (1991) *New Words, New Worlds*. Lampeter, Social and Cultural Study Group, Institute of British Geographers.

Relph, E. (1981) *Rational Landscapes and Humanistic Geography*. Croom Helm, London.

Rogers, C. (1957) 'The necessary and sufficient conditions of therapeutic personality change', *Journal of Consulting Psychology* 21: 95-103.

Rogers, C. (1980) *A Way of Being*. Boston, Houghton Mifflin.

Seamon, D. (1982) 'The phenomenological contribution to environmental psychology', *Journal of Environmental Psychology* 2: 119-140.

Seamon, D. (1984) 'Emotional experience of the environment', *American Behavioural Scientist* 27(6): 757-770.

Seamon, D. and Maugeraurer, R. (1989) (eds) *Dwelling, place and environment: towards a phenomenology of person and world*. New York: Columbia University Press.

Thrift, N. (1996) *Spatial Formations*. Sage, London.

Thrift, N. (1999) Steps to an ecology of place. In Massey, D., Allen, J. and Sarre, P. (eds) *Human Geography Today*. Cambridge: Polity Press, pp. 295-324.

Thrift, N. (2004) Intensities of Feeling: Towards a Spatial Politics of Affect. *Geografiska Annaler* 86(1): 57-78.

Tuan, Y.F. (1976) Humanistic Geography. *Annals of the Association of American Geographers* 66: 266-76.

Tuan, Y.F. (1977) *Space and place: the perspective of experience*. London, Edward Arnold.

Twigg, J. (2000) *Bathing: the body and community care*. London: Routledge.

Tyson, M. (1998) *The Healing Landscape: Therapeutic Outdoor Environments*. McGraw Hill.

Williams, A. (ed.) (1999) *Therapeutic Landscapes: The Dynamic Between Place and Wellness*. University Press of America, Lanham, MD.

Chapter 9

The Geographies of 'Going Out': Emotion and Embodiment in the Evening Economy

Phil Hubbard

Introduction

> The government is keen to generate an urban renaissance where more people live and work in town and city centres. Evening activities are a fundamental part of the urban renaissance because they extend the vitality of a town or city beyond normal working hours ... If the urban renaissance is to be successful, a wider cross-section of people must be attracted into town and city centres in the evening and at night (ODPM, 2003, 3-4).

Much in vogue among politicians, policy-makers and planners – as well as a number of urban geographers – the notion of an urban renaissance has recently been mooted as a solution to the disinvestment and decline that came to characterise many British city centres in the 1980s and 1990s. In simple terms, this putative urban renaissance is about encouraging an increasing number of citizens to live and work in revitalised city centres that offer a diversity of leisure, retail and employment opportunities. Fundamental to this vision of renaissance – and enshrined in the urban white paper '*Our towns and cities: the future*' (ODPM, 2000) – is the idea that the presence of more residents in towns and city centres will increase the viability of the night-time economy, and, conversely, that increasing the vitality of evening leisure will encourage more people to live in city centres. This concept of urban renaissance arguably has its roots in continental Europe where city centre living has remained the norm, and where leisure, business and residential land-use seemingly co-exist in harmony on a twenty-four hour basis. Aping this continental model, British cities including Leeds, Nottingham and Manchester have revitalised their city centres through the careful cultivation of nightlife, providing proof-positive that enhancing the level of activity associated with the city at night results in considerable benefits not just for city centre businesses, but for the entire city (Lovatt, 1997). Accordingly, many other British local authorities have been vigorously pursuing policies designed to reinvent city centres as spaces of leisured consumption, with town centre management teams and business improvement consortia having been 'formed to entice consumers into city centres through a series of entrepreneurial endeavours'. Stressing that city centres offer round-the-clock consumption opportunities, a common goal has been to attract consumers into pubs, restaurants and clubs to extend the 'vitality and viability' of the city centre into the evening (i.e. from 5pm to 8pm) and late-night periods (i.e. after 8pm).

Yet major concerns have been raised about the sustainability of this as a model of urban regeneration given the potential conflicts between city centre residents and those drawn into city centres for leisure and recreation. Here, even promoters of the evening economy concede that careful management is needed to ensure that late-night noise, littering and street urination do not impinge on the residential amenity of those who have often paid a premium to be part of a much-vaunted civic renaissance (Jones *et al.*, 2003). However, alongside these very real concerns there exists another, potentially more damning, critique of policies designed to revive the contemporary evening economy: namely, that they fail to draw a sufficiently wide range of consumers into the city centre, and hence provide an impoverished base on which to build a 'true' urban renaissance. Indeed, a number of academic studies have suggested that efforts to revitalise the evening economy have disproportionate appeal for particular social factions – not least those with high disposable incomes and few family commitments. For instance, Chatterton and Hollands (2002) have charted a steady growth in an alcohol-based, club-oriented culture catering for the 16–25 year-old student and 'youth' market, while Bromley and Thomas (2000) have documented the dominance of younger consumers in Cardiff and Swansea city centres, particularly in the late-night period. Given this form of night-life is seen to have little appeal for many older consumers, in many accounts it is speculated that the revitalisation of the night-time city is actually discouraging some (older) groups from visiting the city centre (Bromley and Thomas, 2000; Hobbs *et al.*, 2000). Some even suggest the devotion of town and city centres to youth-oriented consumption is a significant factor in creating a substantial 'domocentric' population that never goes out because of fears of night-time 'yobbery' and disorderly behaviour (Sparks *et al.*, 2000).

As such, it appears dangerous to suggest that the promotion of a 'continental-style' evening and night-time economy offers a sustainable solution to the problems evident in many British town and city centres. Furthermore, no matter how much planners and policy-makers promote city centre leisure, there is clearly much competition from other sites and spaces for consumer spending. Indeed, recent decades have witnessed massive (corporate) investment in leisure facilities beyond the city centre. In Britain this includes the out-of-town and off-centre retail and leisure parks that arose chiefly in the 1980s and 1990s, including purpose-built developments that provide a range of leisure attractions on a single site adjacent to plentiful and free parking. Typically anchored by a multiplex cinema, such developments characteristically incorporate restaurants, bowling alleys, fast food outlets, and sometimes fitness centres, bingo halls, retail outlets and amusement arcades (Mintel, 1999). By 2001, there were 107 of these 'multi leisure parks' in the UK, taking around 90 per cent of their revenue after 5pm (Jones and Hillier, 2002). Though subsequently discouraged by British planning policy guidance, such multi leisure parks continue to proliferate, and offer a significant alternative to city centre leisure both in Britain and most of Western Europe (Mintel, 1999).

Clearly, the increasing range of opportunities for night-time leisure to be found away from town and city centres means that efforts to promote urban renaissance are fighting against the centrifugal appeal of peripheral sites of urban leisure. Despite this, remarkably little has been written about these new spaces of evening and night-time leisure (though see Jancovich *et al.*, 2002). Indeed, there are precious few academic studies that explore the relative appeal of both in-town and out-of-town leisure for a range of different consumers, with many accounts lapsing into stereotypes of 'lager louts' driving others away from city centres. In this chapter I wish to counter (or at least flesh out) such stereotypes by summarising a research project on the social geographies of night-life in Leicester (UK). A city of some 340,000 inhabitants, Leicester boasts a range of out-of-town leisure spaces (including peripheral multi-leisure parks, multiplex cinemas, and retail parks) as well as a city centre that is currently home to many bars, clubs and restaurants, including several pub chains targeting student and young adult consumers. Against the ongoing efforts of Leicester City Promotions to revitalise and diversify Leicester city centre's evening economy, this project set out to explore the ways that different consumers used both the city centre and out-of-town leisure parks as sites of night-time leisure. To these ends, data on evening leisure routines were obtained through a questionnaire distributed to 1200 households in six neighbourhoods broadly representative of the city's population (these were also located at different distances from the city centre; something often hypothesised as significant in influencing the relative appeal of in-town and out-of-town leisure). 459 questionnaires were returned, providing a valuable insight into variations in leisure behaviour across areas and social groups. From these returns, five individuals from each neighbourhood were selected for semi-structured interviews of between forty minutes and two hours' duration (i.e. thirty interviews in total). This second stage of research focused on experiences of 'going out', with interviewees prompted to describe their last night out and to reflect on both the negative and positive aspects of those experiences. Given the aims of this chapter, and the fact that the results of the extensive survey have been discussed elsewhere (Hubbard, 2002), the remainder of this chapter draws primarily on taped, transcribed and anonymised interview data to elucidate the *emotional* geographies of going out. However, before discussing this rich source of data it is important to explain why a focus on emotions is crucial for comprehending these geographies of night-life – and to consider the limitations of interview data for eliciting these emotional geographies.

Mapping Emotions

As an attempt to escape from the pressures of work and domestic reproduction, people's decisions to participate in particular leisure activities can be understood to have both rational and emotional dimensions (Rojek, 1995). While it is tempting to suggest that such decisions are rational ones about the type of emotional experiences that might be associated with particular activities, the reality is a good deal more complex, with even so-called rational decisions imbued with complex emotional cadence. This is especially true, perhaps, of decisions to 'go out' at night. While much has been made (in geography and elsewhere) of the flattening and homogenisation of space and time, there can be little doubt that night-time remains very different from daytime, being a realm indelibly associated with both the dreams and nightmares of urban living (Schlör, 1998). Indeed, contemporary representations of the city frequently emphasise the forms of *excess* that occur under the cover of darkness, with the British media having recently whipped up moral panics about, amongst other things, binge-drinking, drug-fuelled dance cultures and forms of street violence that are regarded as endemic in the night-time city (and depicted as particularly prevalent in city centres). This combination of the ludic and liminal hints at the emotional extremes that might be experienced by those participating in the evening economy, where anger, elation, sorrow, fear and desire may be routinely experienced (and perhaps all in the space of one evening out).

For such reasons, negotiating the city is a very different proposition in the hours of darkness, potentially involving heightened sensations of anxiety and excitement. For some, 'going out' in the city at night induces anxiety-like symptoms and uncertainty, and is something they prefer not to do (Sparks *et al.*, 2000). Accordingly, 24 per cent of British Crime Survey (2000) respondents say they do not go out at night and another 16 per cent report going out at night only once a month or less. For 57 per cent of these respondents, this was because they felt they had no need to go out, but for 19 per cent fear of crime was a major factor. For others, however, the apparently risky nature of night-life is part of its appeal, and some may actively revel in encounters with difference or immersion in potentially risky or unpredictable situations. For most though the challenge is to negotiate these pleasures and dangers, using practical knowledge of the city to avoid situations that they would rather not deal with while seeking out forms of pleasure and stimulation.

Given cities at night are depicted as replete with dangers and unpleasant experiences – as well as playful and pleasurable ones – exploring how 'going out' produces emotions such as frustration, resentment, pleasure, desire, anger, happiness, fear and anxiety thus appears vital if we are to make sense of the appeal of different sites (particularly the differences between in-town and out-of-town leisure). However, despite geographers' current interest in emotions, there remains much disagreement as to which methods might allow us to best explore people's emotionally-charged experience of the city. For instance, mental mapping techniques provide a way of exploring how individuals cognitively map their preferences onto the cityscape (see Holloway and Hubbard, 2001), yet have been oft-criticised for failing to register the depth and range of emotional experiences associated with different spaces. Related projective methods – particularly semantic differential techniques – can likewise be used to make generalisations about the perceived quality of different areas. However, such behavioural methods have often been accused of downplaying the innate creativity of human agents through the adoption of positivist quantitative methods (see Gold, 1992). In contrast, humanistic methods typically aim to excavate the rich sediment of human emotion that accrues in different contexts by examining a variety of written and spoken 'texts' (e.g. poems, paintings and novels). Richly evocative, these often highlight the bonds of attachment that unite people and place, while associated methods of ethnography and unstructured interviewing can highlight the desires and dreads experienced in different places.

Distinguishing between *topophobia* and *topophilia*, humanistic methods thus offer a viable way of mapping emotional landscapes. However, recent theoretical debates in human geography have raised serious questions about the ability of such approaches to grasp one of the fundamental aspects of 'being in the world': embodiment. In many ways, it seems self-evident that any understanding of emotional attachments to place must take into account the capacities and meanings of the body. For example, it is obvious that access to certain spaces may be constrained or enabled by corporeality, with bodily stature, muscularity and flexibility all influencing people's ability to occupy different spaces (something highlighted in the literatures on geographies of disability and aging – see Hepworth and Milligan *et al.*, this volume). Likewise, it is apparent that the inscription of bodies with gendered, classed, aged and sexed meanings shapes the relation between people and place in powerful ways (so that, for instance, some bodies may be coded as 'out of place' in certain sites). And yet it is only in the last decade or so that geographers have explicitly considered the social meanings of the body alongside bio-medical understandings of corporeality, acknowledging both the plasticity and variability of the body as well as the stubborn recalcitrance of flesh and blood (see Hubbard *et al.*, 2002).

This embodied ontology stresses that individuals are only able to express themselves in space through their bodies – corporeal physicality representing the basis of 'being in the world'. Of course, many of these ideas are not new, being presaged in the work of authors as varied as Mauss, Merleau-Ponty, Goffman and Bourdieu (for an overview, see Hubbard *et al.*, 2002). Rejecting the traditional conception of a centred cognitive being, such 'embodied ontologies' interpret space as more than contextual: it is instead regarded as material the body engages and works *with* (Lupton, 1998). In this sense, emotions can be conceptualised as the felt and sensed reactions that arise in the midst of the (inter)corporal exchange between self and world. Emotions can therefore be regarded as quite distinct from the long-term attitudes, feelings or preferences we express about our environment. As such, while emotions are both a state of mind and a physical experience, it is clear that particular encounters between self and world elicit a strong affective reaction which is emergent rather than pre-given (such as the embarrassment we experience when we arrive inappropriately dressed at an interview or presentation, or the frustration we often experience when interacting with technologies that refuse to work for us). Conversely, managing these emotions is part of the process by which we construct our sense of self, with socio-cultural circumstances dictating that particular forms of emotional management are appropriate for different social groups. For example, it has often been noted that men and women manage their emotional selves very differently, with men encouraged to repress particular emotions that are associated with vulnerability rather than strength

(Williams, 2000). Inevitably, this means men and women might react differently in similar situations, with men typically and characteristically venting their anger, frustration or desire in different ways than women, and expressing this through different body language. This gendering of emotion is often commented upon (see Hochschild, 1983) yet age, class and ethnic identities are also constructed through forms of emotional management and by working though emotional conflicts in different ways.

The way that emotions arise in particular encounters with place has been vividly evoked in the work of Ben Malbon (1999) on spaces of clubbing. Using ethnographic and interview techniques, Malbon reports on the haptic and sensory pleasures of dancing, showing that the material surroundings of a club heighten desire through the combination of lighting, music, décor and the cultivation of a sexually-charged ambience. For those seduced by the spaces of clubbing, the experience is reported to be both affectively intense and emotionally-rewarding. Somewhat differently, Robyn Longhurst (1998) has explored the bodyspaces of pregnant women. Highlighting both the bodily discomforts and social stigmas faced by such women as they go about their day-to-day activities, Longhurst suggests that many pregnant women experience profound anxiety about being in spaces of leisure and consumption that they routinely visited before they were pregnant. As in the work of Davidson (2001a), the suggestion here is that space 'presses upon bodies differently', with spaces that alienate and alarm some being a fundamental locus of others' self-identity (i.e. a place in which they feel at ease and untroubled). Davidson's (2000) work is exemplary in many other respects, illustrating some of the strategies that agoraphobics employ to 'keep a hold of themselves' in spaces where the boundaries of the embodied self are difficult to define. Using qualitative methods to explore what are inevitably highly sensitive and personal issues, Davidson's use of experiential accounts demonstrates that spatialised narratives can reveal many of the emotions experienced as emergent senses of the embodied self are constructed in relation to people's surroundings.

Nonetheless, talking seems to have some obvious limitations as a way of exploring people's emotional experiences of different spaces. After all, emotions are embodied, in the sense that they are felt and experienced, and it might be suggested that language cannot adequately capture the way emotions are felt (e.g. the flush of embarrassment, the rise of anger, the pang of disappointment). Furthermore, others might contend that emotions arise from an inner (unconscious) self, and thus cannot be consciously articulated. However, and following Mehta and Bondi (1999), this project started from the assumption that analysing 'emotional talk' is a useful way of interrogating the emotional topographies of the night-time city. This type of approach explicitly rejects the notion that emotions cannot be articulated because they dwell within the psyche (cf. Pile 1996). However, it does not imply any straightforward correspondence between individual emotional experience and its articulation. Instead, it interprets emotions as primarily interpersonal phenomena (i.e. arising in response to social circumstances), stressing that the public representation of emotional experience is not secondary to its private experience. Parkinson (1998) stresses that discourse is always and inevitably emotionally-charged, and thus reveals responses to specific social situations: like their management of emotion, people talk about their emotions according to their position in interpersonal social networks. As Davidson (2000, 32) argues, this means that to successfully chart emotional experiences it is necessary to be critical and reflexive about the act of interviewing, and pay due care to the dynamic relationship of interviewer and interviewee.

Emotion in Spaces of Night-life

This research project accordingly began from the standpoint that the capacity of urban leisure spaces to appeal to different consumers can only be understood if we consider emotional experiences of these spaces. Talking of their experiences of evening leisure in particular spaces, interviewees thus described a diversity of interactions that had elicited negative emotions (e.g. fear, disgust, anger, frustration) and those that excited more positive emotions (e.g. love, happiness and desire). Here – and perhaps unexpectedly – it was clear that the city centre and out-of-town sites were associated with very different affective intensities and emotional experiences. Taken together, the findings suggested that multi leisure parks and other out-of-town spaces in Leicester were sites where it was considered easy to remain in control, comfortable and at ease, yet where it was still possible to experience a variety of pleasures. In contrast, the city centre at night was perceived to constitute a more 'difficult' and unpredictable space, where the boundaries of the embodied self were more regularly brought into question – in both desirable and undesirable ways.

   In this sense, it can be argued that the decision to go out-of-town or visit the city centre was profoundly shaped by individuals' understandings of the type of emotional management that might be required in a given setting. As such, Leicester's multi-leisure parks were popular with particular consumers at particular times because they were seen to facilitate strategies of emotional management that they regarded as crucial in negotiating the contemporary night-time city. In contrast, the design and production of city centre leisure space was seen to require a different form of resourcefulness, and one that meant its appeal was both selective and variable at different times of the week. In the following discussion, some of the key strategies of emotional management that people utilised as they participated in the night-time economy will be discussed, with particular reference being made to the ways that these were enabled or problematised in out-of-town and city centre venues respectively.

Peripheral Leisure Venues: A Quiet Night Out?

Goffman's classic accounts of public life, and especially *The Presentation of Self in Everyday Life* (Goffman, 1980), demonstrate that a particularly common strategy for retaining control of self in social spaces is to affect indifference to others. As he suggests, when in public spaces we therefore often adopt modes of civil inattention to those strangers who surround us, granting them their own 'personal space' while we seek to maintain ours. This means that social interaction is primarily restricted to a *visual* exchange, with a concomitant respect for personal space and the maintenance of a neutral (or at least unthreatening) demeanour towards others. This form of inattention is particularly manifest in the way that people scan each other visually to ascertain they do not pose any threat, without overtly staring at them (i.e. *uncivil attention*). This ability to move through social space without attracting undue attention is one associated with certain mythical figures of the urban scene – such as the streetwise *flaneur* – but Goffman underlines that this ability to limit involvement in social space is one that is, to a lesser or greater extent, employed by all urban citizens (and one that Anderson, 1990, suggests is increasingly crucial in avoiding danger). However, this ability comes easier to some and is easier in some places than others.

Interviews with consumers in Leicester suggested that part of the appeal of out-of-town leisure spaces was the fact that they were seen to facilitate such modes of civil inattention, and thus enabled people to enjoy 'themselves' (and those they had chosen to go out with) without experiencing anxiety about the boundaries of self being brought into question. This does not imply that leisure in these spaces is passive, as people perform their presence in these spaces through embodied practices that involve an active engagement with their surroundings (e.g. choosing a parking space, parking their car, ordering food, watching a film, playing a videogame and so on). However, it does suggest that the social interaction with strangers characteristic of these out-of-town leisure parks is essentially limited. Bauman's description of 'public non-civil spaces' thus seems particularly appropriate:

> Such places encourage action, not interaction ... The task is consumption, and consumption is an utterly, irremediably individual sensation. The crowds filling cathedrals of consumption are gatherings, not congregations, clusters, not squads, aggregates, not totalities. However crowded they might be, there is nothing collective in these places (Bauman, 2000: no pagination).

Hence, one of the key social roles of the out-of-town leisure parks is offering an *affective ambience* that offers the illusion of social mixing but where there is little need to interact with unknown others or negotiate the boundaries of self.

This tendency can be illustrated with reference to Meridian Leisure Park, a cinema, health club and restaurant complex located three miles to the south east of Leicester city centre. Though no interviewees claimed to have ever gone to this park alone, it was also notable that few interviewees reported interacting with other consumers in anything but a superficial or visual manner. As one respondent reported:

> It's a place I usually go with a group of friends from work. We generally go straight from work, taking it in turns to drive. It's about once a month, a trip out for a meal and then to see a film afterwards. I guess it sounds pretty boring, but it's usually the same group of us ... it gives us a good chance to catch up on gossip, have a bitch about work, you know? It's not like a big night out where you're expecting to pull or anything [*laughing*] (Katy).

The fact that most visitors to the Meridian Leisure Park did so in well-defined groups (most frequently, as a group of friends or a 'couple') and do not seek chance encounters, thus predisposes them to 'desocialise' the space around them (Davidson, 2003). This involves consumers presenting themselves in a non-threatening and relatively inconspicuous manner that does not invite comment, with respondents stressing they did not 'dress-up' when visiting this site, nor wear their work clothes, but rather adopted a 'casual' look: 'dressing-down, jeans and a t-shirt' (John) according to one, 'pretty much what I wear at home anyway, a long skirt, a jumper' (Rosie), according to another. This bodily presentation often extended to the adoption of 'tie signs' that indicate that someone was in a pre-defined group, such as holding hands with a partner. (Observational work at the same location revealed a variety of familiar strategies employed by people standing alone while waiting for friends, such as constantly glancing at their watch – a tactic commonly used to prevent the intrusion of others into one's personal space.)

By sending out signals that they wish to maintain their privacy, most groups thus interact with other groups only in the visual sense. Associated with this is a concomitant maintenance of a 'social distance' between groups, with few consumers seeking to speak with others or impinge on their 'personal space'. In many ways, the design of out-of-town leisure parks facilitates such strategies. To illustrate this we might reflect on the design of multiplex cinemas (which are an integral part of most leisure parks, and the main attraction at the Meridian Leisure Park): traditionally, cinemas provided large foyers where people could mingle after films to discuss the film they had just seen (or were about to see), or make plans to go on elsewhere. At Meridian Leisure Park, however, the foyer of the cinema is dedicated to the selling of concessions, and while it may be themed around readily-identifiable cultural symbols (such as cartoon characters), it is not designed to encourage interaction between people. Moreover, with the multi-screening of films, the foyer has taken on the characteristics of a 'space of flow' rather than a public place: it is carefully designed to ready the audience for the visual pleasures of cinema (as well as reminding them of the 'guilty' pleasures of consuming popcorn, soft drinks and sweets [see Germann Molz, this volume]), but not to encourage mixing between different groups. For Fiona, it was important that the Warner Village cinema at Meridian Leisure Park allowed for purchasing tickets quickly, so there was little need to 'hang around' in the foyer:

> You walk in and there's like ... a few booths, four booths or something and there's a definite queuing system to get into the booths, so that's one thing that's quite good – knowing where you're going, and how to pay and where to pay ... (Fiona).

Anxieties about queuing at cinemas were reiterated by the other respondents, with Tracey detailing how she got embarrassed when her children had got bored waiting in line and started running around. For most respondents, the foyer was accordingly not regarded as a social space, but a space to pass through as quickly and easily as possible: there was little that encouraged them to dwell. This configuration of bodies and objects is carried through into the auditorium, where it is common practice for consumers to choose seats away from others. Given most films in Britain are shown in auditoria where occupancy rates are less than 30 per cent at any given time (Mintel, 2000) it is usually not difficult to maintain a degree of personal space by sitting away from others. Fiona encapsulated the views of many when she argued that 'We always look for a seat in an empty row ... you don't want some noisy kid next to you do you? Or someone who wants to strike up a conversation in the middle of the film'.

The strategies of managing personal and social space characteristic of multiplex cinemas are, to a lesser or greater degree, replicated in other peripheral leisure spaces in Leicester, from fast food drive-thrus to health clubs. In each case, consumers' desire to avoid unwelcome attention was facilitated by the design and layout of these spaces, which environmental psychologists might well characterise as 'sociofugal' (Holloway and Hubbard, 2000). Here, fears about bodily defilement or disturbance appeared to be significant factors predisposing some consumers to visit out-of-town sites rather than in-town spaces of leisure. By providing an environment where civil inattention is the norm, out-of-town leisure parks apparently made it easy for consumers to maintain a strong sense of self and to avoid any anxiety about the boundaries of their body being defiled, dirtied or brought into question (see also Sibley, 1995, on boundary-drawing). Interviewees accordingly stressed that they felt more 'relaxed', 'comfortable' or 'off-guard' at multi-leisure parks than in the city centre. Interestingly, many of these had been frequent users of the city centre a few years earlier, but claimed variously that they were 'getting too old for that kind of thing' (Peter) or simply that the city centre had become more threatening:

> The clubs I use to go to are still there but they've changed names, and, you just felt safer then than now, and it was better. The nightclubs were better, you had better times than you do now, because you've got to watch yourself all the while when you're out (Tracey).

For many such consumers, a visit to an out-of-town leisure park had become their 'big' night out, an occasion reserved for the week-end and typically involving a meal and a drink in an adjacent restaurant before a visit to the cinema, or perhaps a take-away afterwards. However, while some other interviewees similarly regarded out-of-town sites as places where it was easy to be relaxed, they suggested that they lacked the 'buzz' or 'excitement' of the city centre, which they regarded as more attractive for a 'big' night out. For some, therefore, the more unpredictable – and perhaps less restrained – social relations typical of city centre leisure spaces continued to hold considerable appeal.

In-town Leisure: The Pleasures and Perils of the 'Big Night Out'

In contrast to out-of-town sites of night-time leisure, which, to quote Bauman (1993) are essentially 'desocialised', the city centre was widely perceived and experienced by interviewees as a social *meeting place* – that is to say, a site characterised by more varied and often less civil encounters than those associated with out-of-town venues. Hence, while multi-leisure parks like Meridian were perceived as facilitating civil inattention and indifference to others, the city centre was often reported as a setting where personal space was repeatedly and routinely bought into question in a variety of ways. For some, this was an obvious source of stress and anxiety:

> There's a whole load of new bars on Granby Street [*central Leicester*], so we went a few months ago to give them a whirl ... [*But*] they were so packed. Nowhere to sit down, long queues at all the bars, too much hassle. And like all the pubs in the centre, really loud music, couldn't hear ourselves think (John).

This tendency to essentialise city centre venues as exuberant, lively and crowded was widespread, with many interviewees glossing over the pleasurable experiences of collective consumption to outline the negative consequences of this:

> There's this bar I go to at lunchtime, and they do OK food and at a reasonable price. So we thought we'd do some shopping after work and then go there for a drink and a bite to eat. Big mistake ... It was impossible for me to get to the disabled toilets without getting loads of people to move, and I really hate that, especially when everybody's trying too hard to help. Awful (Mary).

This change in character of the city centre between day-time and night-time was referred to by many, with several describing their apprehension about going into venues or streets that they had often visited during the day, but were less familiar with at night.

For some, this apprehension was connected to their perception that city centres are host to a wider and more diverse group of users than peripheral leisure parks (something not borne out by survey data, which suggested that the range and diversity of social groups at out-of-town sites was, if anything, wider than that associated with the city centre). As such, several interviewees reported anxieties that appeared to be connected to their fears of social difference. Fiona, for example, stressed:

> I used to go to the Odeon [*city centre cinema*] ... but you got too many kids hanging around, being noisy, skateboarding and that. And smoking too, in the foyer and once in the cinema ... they didn't stop when I asked. I get asthma so it really bugged me. You just don't get that type of crowd at the Meridian [leisure park] ... (Fiona).

Likewise, Perzana stated 'If you go to the Odeon, there's very big groups and very laddish, rowdy'. Another, Mark, claimed that he wouldn't go to the city centre on a weekend because the dominant social group changed: 'It's full of what I would call "townies", the beer and kebab crowd ... Trying to get a taxi on a Friday night is dicing with death'. This latter quote illustrates a crucial point: namely that concerns about encounters with social difference were time- as well as space-specific, with several respondents claiming that while they had enjoyed nights out in Leicester during the week, they had witnessed some unpleasant and disturbing scenes at weekends.

This concern was echoed in interviewees' stories of specific times they had felt fearful in Leicester city centre. For many, such instances occurred where they had been walking through parts of the city and encountered people who, for a variety of reasons, were not conforming to the modes of civil inattention that might normally be associated with the streets during daytime. These ranged from relatively 'harmless' yet temporarily unsettling encounters with drunken individuals to instances of unwelcome sexual attention and, in one case, a street robbery where the victim sustained a broken arm and severe bruising. Even in the context of a lengthy and relaxed interview, encouraging people to recount these experiences was difficult, and in any case, the unwillingness of people to elaborate on the details of an incident reveals much about their own ways of managing emotion. What was clear, however, was that these had led to the adoption of a variety of 'coping strategies' by which the victims sought to deal with their trauma. One respondent, for instance, confronted his fears by returning to the location of an incident the next week to reassure himself that the incident had been a 'one-off'. More typically, however, people sought to avoid such unpleasant or unsettling incidents by avoiding the area, or using a taxi or car rather than walking around town.

The fact that using a car allowed interviewees to feel more secure and comfortable as they engaged with the city at night is a significant finding, and one that helps shed light on the use of different leisure sites. To date, social geography has downplayed the significance of automobility, which creates distinct ways of dwelling, travelling and socialising within cities. Several sociologists – most notably Urry (2001) – offer a valuable corrective, theorising automobility as the dominant culture that organises and legitimates socialities across different genders, classes and ages. Following Urry, it appears important to consider the way that automobility shapes involvement in the night-time economy, necessitating thinking about the pleasures and dangers of driving through the city. While those who claimed they would normally walk or rely on public transport when they went out in the evening reported avoiding certain routes at certain times, in contrast, car owners reported little anxiety about driving through any areas of Leicester at night. Most appeared entirely happy to drive through an area they would not walk through at night; the car seemingly provided them with ontological security, regarded as a cocoon that surrounds like a second skin (Lupton, 1999), as this quote stresses:

> Probably quite stupidly, I feel quite safe within my car. I often lock the doors when I'm driving on my own at night, or when I drive home from [*work*] ... it's not particularly nice round Leicester at night, I'd always lock my doors, but otherwise it doesn't bother me (Fiona).

However, respondents did report anxiety about leaving their car in the city centre, with several reporting how upset or distraught they had been in instances when their car had been damaged or stolen in the city centre. In contrast, out-of-town leisure venues were regarded as 'secure' by Ian:

Meridian Park is much safer for parking than in Leicester. There always seem to be people around ... I think that busy car parks are much safer than ones where there are only a few cars around. There are a lot of car parks in Leicester, but some of them feel a bit isolated at night, off the beaten track (Ian).

The view that out-of-town parking is inherently safer than city centre parking was also noted by Thomas and Bromley (2000), who found that 89 per cent of night-time users of Swansea were uneasy about the safety of their cars. For many, fear of car crime is as pressing a concern as fear for their own personal safety; they regard the car as an extension of self, and the threat that this personal space may be broken into shapes their decision to use 'safer' out-of-town sites.

This anxiety about parking in the city centre, coupled with fears about the range of unanticipated encounters that might occur between car park and leisure venue, thus predisposed some interviewees to avoid the city centre all together. Coupled with this is the pleasure that some take in driving, as one interviewee explained with reference to his decision to go to an out-of-town restaurant rather than a similar one located in the city centre:

I just enjoy driving, you know. Not sitting in traffic, like during the day, but just enjoying a drive ... getting away, feeling free to go where I want (John).

Another explained why driving out of the city might be preferable to a frustrating journey into town:

It's further than the Odeon, but you never can be sure about the traffic when you go into town, you have to time it right. The Warner Village is further but there's always plenty of parking and it's a more predictable journey, you know? It's like, you go into Leicester to see a film or go to the Haymarket [*theatre*] you have to fight the traffic, find a parking space ... Going to Meridian you can time it much better, which is really important if you've got screaming kids to look after (Tracey).

Such assertions concur with the assumption that peripheral leisure parks are frequented by car-borne 'parkaholics' who enjoy the 'out-of-town' experience (Mintel, 1999). Indeed, although the Meridian Leisure Park in Leicester was served by public transport, and was only a short taxi journey for some respondents, all interviewees who reported visiting Meridian on their last evening out did so by private car. Respondents explained this with reference to the fact that it entailed an easy journey 'out' of the city to a one-stop leisure location, where facilities were easily accessible from a car park where they knew there would be spaces. Rather than detailing the exclusion of non-car owners from peripheral spaces of leisure, this study thus points to the enhanced mobility of car owners (see also Beckmann, 2000).

The idea that those who are most mobile avoid city centres in favour of out-of-town venues concurs with Hannigan's (1998, 73) notion that consumers increasingly seek leisure that offers escape from everyday routines while providing 'recurrence of reassurance'. Yet this wish to avoid the unknown is not universal, with Hannigan also noting the existence of consumers who continue to take pleasure in encountering difference, engaging with the complexity of the city, and socialising with strangers. In the Leicester study, such tendencies were most apparent among those consumers who tended to frequent city centre venues, pubs and restaurants, particularly at weekends. Unlike a 'night out' at a peripheral leisure park, which appeared likely to be planned in many respects, most of those going out in the city centre reported having little set idea of where they might be going in an evening, how much they were going to spend well in advance of their visit or even how they were going to get home:

> I normally e-mail or text a few friends in the afternoon, arrange to meet up in a pub about eight, and we take it from there ... Where we end up depends on who turns up: might be a club or sometimes we might go for a meal after the pub ... There's always new places to go, so I don't think I could describe a typical night out. But that's what makes it good, having a choice (Sarah).

This notion of spontaneity and 'going with the flow' was seen as part of the appeal of the city centre for some, given it seemed to offer a range of possibilities foreclosed by the limited number of venues at out-of-town sites:

> Start off at the Orange Tree [*public house*] usually, take it from there. There's always a good crowd in there, you know, not just students and that, but a good laugh. We might hook up with some people there, go with them to somewhere they know ... a new bar or something. There's been lots of times we've ended up at a party somewhere, I've had no idea where we are or how we ended up there ... [*laughs*] (Steph).

The idea that abandoning one's inhibitions, letting down one's guard (often with the help of alcohol) and losing one's self in the city are highly enjoyable experiences has rarely been tackled by geographers, perhaps conscious of the difficulty of charting 'intoxicated' geographies. Yet in this study, as in the work of Chatterton and Hollands (2002) there was much evidence of the varied reasons that people engaged in nightlife, with 'traditional' concerns such as courtship and kinship interlaced with pleasures of consuming music, alcohol, and drugs in venues designed to heighten these affective intensities in a variety of strong ways (see also Malbon, 1999).

In sum, it appeared that the city centre and out-of-town were associated with markedly different (and in some senses antithetical) forms of emotional management. For most interviewees, the chief appeal of out-of-town sites appeared to be their relaxed and casual ambience, which allowed people to adopt modes of civil inattention that were not so different than those associated with sites of consumption and sociality during the daytime. In contrast, the city centre at night was experienced by many as more unpredictable and perhaps unknowable. The antithetical emotions this elicited among different interviewees (e.g. excitement and fear) points to the varied ability of different groups not just to access different sites of leisure, but to enjoy using them as sites of leisure and recreation. Many interviewed here simply found the city centre too stressful or difficult to cope with, and hence shunned it all together. Yet the converse was not the case, with those interviewees who socialised in the city centre also claiming to enjoy out-of-town leisure. Indeed, while no respondents reported combining a visit to an out-of-town leisure park with a trip to the city centre on a given evening, many respondents used both at different times, and developed more or less varied leisure routines where both types of space fulfilled a particular emotional need (for instance, it was relatively common for younger consumers to go out-of-town during the week, but reserve a 'big night-out' in Leicester city centre for the weekend, while for many older consumers a trip out to a leisure park constituted their 'big night out'). The fact that Leicester's out-of-town leisure spaces appear to have appeal for a wider range of consumers than the city centre raises some important questions about promoting urban renaissance in the city. How these challenges might be best met without eviscerating city centre nightlife and damaging its appeal for those who thrive on its edgy nature is a vexed problem that Leicester city council, as well as others in the UK, are currently seeking to answer (see ODPM, 2003).

Conclusions

In contrast to the daytime city, the city at night is often represented as a space of extreme emotional reactions (Schlör, 1998). Interactions and encounters that provoke little unease during the day might therefore become a source of anxiety at night, while, conversely, the darkness of the city at night seems to heighten the excitement, pleasure and enjoyment experienced in many spaces (not least spaces of consumption). The city at night is therefore often described as profoundly *ambivalent* offering both opportunity and threat in equal measure. Focusing on the use of night-time leisure spaces, this chapter has brought this into question, and suggested that this ambivalence plays out very differently for different individuals. For many, peripheral leisure spaces are popular precisely because they allow consumers to avoid many of the urban areas that they associate with negative emotions. Further, this chapter has suggested they are both represented and experienced as 'family' leisure spaces, where family has become a synonym for predictability, polite conduct and 'shallow sociality'. Hence, while the range of attractions on offer at most out-of-town leisure parks are stunningly predictable and sanitised, the appeal of these attractions for many consumers is undeniable; they give them the ability to experience the bodily pleasures of consuming (e.g. bowling, drinking, eating, visually consuming film etc) but in an environment that offers little danger of them experiencing negative emotions (e.g. getting dirty, being jostled in a crowd, being confronted with unpredictable encounters, being subjected to unpleasantly loud noise, experiencing physical and/or emotional discomfort etc). While it might be argued that these emotional geographies of leisure might also be encountered/experienced in specific city centre sites of leisure, the association of Leicester city centre at night with particular types of encounter clearly predisposes many consumers to shun it in favour of peripheral leisure parks. On the other hand, it needs to be stressed there were many respondents in this study who regarded out-of-town leisure parks as predictable, mundane and even sterile, using them only rarely, and typically only mid-week.

Suggesting the decision to go in-town or out-of-town is an emotional one in many senses, this chapter has nonetheless merely scratched the surface of the complex emotional geographies of the city at night. There is clearly much more that could be said about the embodied pleasures of eating, drinking, dancing, talking, driving and otherwise participating in the evening economy, as well as the frustrations, angers and sorrow that might arise (see also Thrift, 2004). In the final analysis, this underlines the pressing need for further affective and emotional geographies that shed light on the complex and contradictory ways that different class, ethnic, gender and age groups negotiate the pleasures and the dangers of the night-time city. The idea that evening and night-time leisure is emotionally-charged has not been widely explored, but offers massive potential for understanding people's participation in an evening economy that is an increasingly important part of the urban economy. Indeed, if one follows Parkinson (1998) in arguing that emotions are *effects* of transactions between people, places and things, it seems essential that geographers investigate the emotional experiences of night-time leisure. While this necessitates an appreciation that identical experiences can engender antithetical emotions in different consumers (e.g. poor service in a restaurant can provoke either resignation or anger), this is not to argue for the revival of humanistic and/or subject-centred approaches where knowing subjects are bequeathed with an endless capacity to shape their material world. Instead, it underlines the need to extend consideration of what is material in studies of 'material culture' to encompass the emotions that are a necessary accompaniment to the interactions between people and place. It is through an exploration of these emotional transactions that I would suggest a more thoroughly *social* geography of the evening economy might be developed.

References

Aitken, S.C. (1998) *Family fantasies and community space* New Brunswick: Rutgers University Press.
Anderson, E. (1990) *Streetwise: race, class and change in an urban community* Chicago: Chicago University Press.
Bauman, Z. (1993) *Postmodern ethics* Oxford: Blackwells.
Bauman, Z. (2000) 'Urban battlefields of time/space wars' *Politologiske Studier* 7: September www.politologiske.dk/artikel01-ps7.htm.
Beckmann, J. (2001) 'Automobility – a social problem and a theoretical concept' *Environment and Planning D – Society and Space* 19: 593-607.
Chatterton, P. and Hollands, R. (2002) 'Theorising urban playscapes: producing, regulating and consuming youthful nightlife city spaces' *Urban Studies* 39: 95-116.
Davidson, J. (2000) '... the word was getting smaller: women, agoraphobia and bodily boundaries' *Area* 32: 31-40.
Davidson, J. (2001a) 'Fear and Trembling in the Mall: Women, Agoraphobia and Body Boundaries', in Isabel Dyck, Nancy Davis Lewis and Sara McLafferty (eds) *Geographies of Women's Health* London and New York: Routledge, 213-230.
Davidson, J. (2003) 'Putting on a face: Sartre, Goffman and agoraphobic anxiety in social space' *Environment and Planning D: Society and Space* 21: 107-122.

Goffman, E. (1980) *The Presentation of Self in Everyday Life* Harmondsworth: Penguin.

Gold, J.R. (1992) 'The decline of cognitive-behaviouralism in geography and grounds for regeneration' *Geoforum* 23 239-247.

Hannigan, J. (1998) *Fantasy city: pleasure and profit in the post-modern metropolis* London: Routledge.

Hochschild, A. (1983) *The Managed Heart: Commercialisation of Human Feeling.* Berkeley: University of California Press.

Hobbs, D., Lister, S., Hadfield, P., Winlow, S. and Hall, S. (2000) 'Receiving shadows: governance and liminality in the night-time economy' *British Journal of Sociology* 51: 701-717.

Hollands, R. and Chatterton, P. (2003) 'Producing nightlife in the new urban entertainment economy: corporatization, branding and market segmentation' *International Journal of Urban and Regional Research* 27: 361-385.

Holloway, L. and Hubbard, P. (2001) *People and Place: the extraordinary geographies of everyday life* Harlow: Prentice Hall.

Hubbard, P., Kitchin, R., Bartley, B. and Fuller, D. (2002) *Thinking geographically: space, theory and contemporary human geography* London: Continuum.

Jancovich, M., Faire, L. and Stubbings, S. (2002) *The place of the audience: cultural geographies of film consumption* London: BFI.

Jones, P. and Hillier, D. (2002) 'Urban leisure complexes in the UK: planning and management issues' *Management Research News* 25: 75-83.

Jones, P., Charlesworth, A., Simms, V., Hillier, D. and Comfort, D. (2003) 'The management challenges of evening and late night economy within town and city centres' *Management Research News* 26: 96-103.

Longhurst, R. (1998) 'Representing shopping centres and bodies: questions of pregnancy', in Ainley, R. (ed.sssssss) *New frontiers of space* London: Routledge.

Lovatt, A. (1997) 'Turning up the lights in the cities of the night' *Planning* 1224: 28.

Lupton, D. (1998) *The emotional self* London: Sage.

Lupton, D. (1999) 'Monsters in metal cocoons: road rage and cyborg bodies' *Body and Society* 5: 57-72.

Malbon, B. (1999) *Clubbing: clubbing cultures and experience* London: Routledge.

Mehta, A. and Bondi, L. (1999) 'Embodied discourse: on gender and fear of violence' *Gender, Place and Culture*, 6: 67-84.

Mintel (1999) Leisure: in-town versus out-of-town, *Leisure Intelligence Reports*, www.mintel.co.uk.

Mintel (2000) The UK cinema market, *Leisure Intelligence Reports* www.mintel.co.uk.

ODPM (2000) *Our towns and cities: the future* London: Office of the Deputy Prime Minister.

ODPM (2003) *The evening economy and the urban renaissance* London: Office of the Deputy Prime Minister.

Parkinson, B. (1998) *Ideas and realities of emotion* London: Routledge.

Pile, S. (1996) *The Body and the City: Psychoanalysis, Space and Subjectivity* London and New York: Routledge.

Rojek, C. (1995) *Decentring leisure* London: Sage.

Schlör, J. (1998) *Nights in the big city* London: Reaktion.

Sibley, D. (1995) *Geographies of exclusion: society and difference in the West* London: Routledge.

Sparks, R., Girling, E. and Loader, I. (2001) 'Fear and everyday urban lives' *Urban Studies* 38: 885-898.

Thomas, C.J. and Bromley, R.D.F. (2000) 'City centre revitalisation: problems of fragmentation and fear in the evening and night-time economy' *Urban Studies* 37: 1403-1429.

Thrift, N. (2004) 'Intensities of feeling: towards a spatial politics of affect' *Geografiska Annaler* 86B: 57-78.

Urry, J. (2001) 'Transports of Delight' *Leisure Studies* 20: 237-245.

Williams, S. (2000) *Emotion and social theory: corporeal reflections on the (ir)rational* London: Routledge.

Chapter 10

Environments of Memory: Home Space, Later Life and Grief

Jenny Hockey, Bridget Penhale and David Sibley

Introduction

In 1997, 83 per cent of all deaths took place among people over the age of 65, yet, as Field (2000) points out, limited academic attention has been given to this most common of deaths. A related area of neglect is the inevitable bereavement of older widows and widowers. Yet as Bennett (1997) notes in one of the few studies in this area, this 'high-probability life event' has been experienced by 36 per cent of the population over 65. While the 'positive ageing' emphasis within the gerontological literature tends to mean that death and bereavement are given low priority, the apparent ordinariness of older adults' deaths and bereavements seems to have made them invisible in the thanatological literature too.

This chapter is situated within these gaps. It draws on a currently important theoretical strand within bereavement studies – memory, memorialisation and material culture – to investigate the experience of losing a heterosexual partner in later life. This focus on memory reflects the ongoing academic and therapeutic interest on continuing bonds between the living and the dead (see Klass *et al.*, 1996; Riches and Dawson, 1998; Hallam and Hockey, 2002), and links with theoretical perspectives developed in the field of material culture. When we ask what the gerontological literature has to say about memory, we find memory treated either as a faculty – which might be fading – or as a therapeutic activity known as 'reminiscence'. While reminiscence is indeed seen as a way for older adults to repair the losses of later life (for example, Coleman, 1994), there seem to be few explicit links made between this and the bereavement literature. Instead it is the therapeutic and practice implications of older adults' memories which authors address (Bornat, 1994). And, although there is recognition that the objects, images and sounds recollected across a lifetime may potentially occupy a special place in older adults' lives, they tend to be treated as passive props or stimuli to more active reminiscence 'work'. Yet, as this chapter argues, we need to ask how such items actually operate. Here we will draw on the notion that objects and spaces have their own agency. They therefore need to be seen as not only located within, but also *constitutive* of the social time and space of later life (see for example, Hallam and Hockey, 2001; Pels, 1998; Spyer, 1998).

The data presented here derive from a study which asked how older people's social lives might be curtailed or expanded with the death of a heterosexual partner, and whether they felt comfortable living alone in a previously shared home. What became apparent, however, was how intimately the practicalities of shopping and socialising, cooking and cleaning, were bound up with the emotional transition of bereavement. The materially-grounded nature of these data proved inseparable from their affective import.

Space, Materiality and Later Life

Issues of space and materiality are indeed evident within the gerontological literature in that historically it has had a strong policy focus. In Thane's view (1983), this orientation reflects the 'problem of the elderly' which emerged for the first time in the 1870s with older adults' gradual age-based 'retirement' from paid work. It bred a subsequent concern with the materialities of their health, income and housing (see Hepworth, this volume), a construction of later life vulnerability which resulted in older adults' bodies rather than emotions being prioritised. As Bytheway notes: 'The literature on the history of old age has tended to revolve around matters of number, household arrangements, poverty and the role of old people in society' (1995:15). Thus, when home spaces are being designed for older adults, the focus tends to be on 'comfort, convenience and safety' (Fairhurst, 1998:98). In response to designers' priorities, Fairhurst suggests that we need to look to the social and emotional aspects of older adults' lives and her research details older people's concerns about the *social* nature of home space – what kind of people live nearby, where can grandchildren sleep over, where can a visitor be invited to sit down, where will furniture and personal memorabilia be put? If these needs are not met then sheltered accommodation remains a *space* for eating and sleeping rather than a *place* to feel 'at home' in. At its bleakest therefore, the design of older people's accommodation abstracts them from their social context of family and friends. Designers' emphasis on picture windows, for example, constructs the elderly person as an isolated observer whose participation in an externalised world is reduced to 'watching the world go by' (Fairhurst, 1998:99).

This chapter builds on Fairhurst's concern with the social dimensions of material space to suggest that not only need *living* family and friends be incorporated into planning decisions, but also the spatio-temporal materiality of social networks which include the dead – particularly deceased spouses who once shared the current home space. Though images of the older adult as a guardian of family memories and mementos are often taken for granted – the parent who hangs on to old school reports, the grandparent who alone can name the people in the sepia photographs – the relationship between loss, later life and material culture merits proper investigation. Failure to undertake this means that the items with which older adults live can be mistaken for quirky vestiges of bygone days. Like the elderly person themselves, supposedly viewing the world through their picture windows, these items can be picturesquely marginalized, at best something to rake over for the antique or bric-a-brac shop, at worst fodder for a car boot or garage sale.

The Research Project

The data discussed here therefore shed light on later life spousal bereavement as a spatialised experience. Qualitative interviews were conducted with twenty older widowed people living in and around a large city in the north of England and its neighbouring market town. Seventeen members of the sample were aged seventy or over with another three being in their late sixties. All of them had been bereaved for at least eight months, sixteen for between one and nine years, with another two widowed for eighteen or more years. Sufficient time had therefore elapsed for changes in their relationship with space to have become evident and in many cases stable. All the interviewees were living alone in the home they had shared with a spouse, half in sheltered housing designed specifically for older people, and half in their own private accommodation. Access to participants was negotiated either through the wardens of older people's sheltered housing schemes, the co-ordinators of bereavement friendship groups and drop-ins, or personal contacts.

What we wanted to know was how older adults' relationship with domestic and public space might have changed with the loss of a heterosexual partner. In their accounts of the difficulties of getting out and about, and the changes in their domestic arrangements and routines, data emerged about how their material environments mediated ongoing social relationships with a deceased partner. That their homes might be a site at which these relationships were lived out was not an element which had been built into this gender oriented study. However, since the interviews were designed as opportunities for older adults to establish their own agendas, we soon became aware of what Appadurai (1986) describes as 'the social life of things', that is, the way *social* environments are produced as a result of the way home space and all that was contained within it had now come to embody the past in the present.

The project's sample design produced data about the different housing experiences of privately owned and sheltered accommodation. These reveal the constraints of sheltered housing, both as a limited space for the display of significant items, and as lacking scope for re-arranging living areas to reflect changed personal circumstances. As Kellaher explains, the amount of space available contributes significantly to individuals' bodily engagement with their material surroundings, or 'embodiment in environment through routine' (2001:223). Her data show that people living in residential homes with rooms of less than $12m^2$ have very restricted scope for storing or displaying memorabilia and tend to gravitate – sometimes reluctantly – to more public rooms to escape their cramped conditions. These data sensitise us to the significance of objects and spaces still available to those who live on in their own homes. Fairhurst (1998) shows sheltered accommodation to be a home space where individuality can become submerged within the social category, 'the elderly', an example of how the 'dependency' of older adults can make them vulnerable to externally imposed social identities (Hockey and James, 2002; Milligan, Bingley and Gatrell, this volume). Accommodation designed specifically for older adults – residential homes, sheltered accommodation – helps impose such identities via the extent and organisation of living space. The data we have gathered within private accommodation, however, show that older adults may manage and maintain personal identity precisely *via* home space. As Kelleher says, private settings allow the possibility that 'older people constantly work at shaping their domestic environments, through micro-improvisations that are both material and cognitive' (2001:219). Home space can therefore act as both a spatial resource *and* constraint. How these different possibilities mediate individuals' emotional lives is core to the investigation discussed here.

Agency and the Inanimate

To sum up so far, this chapter has described a *social* environment which housing policy may ignore, but has also argued that social gerontologists themselves can neglect the social presence of partners now absent in embodied form. And to extend this point, it has also been noted that social scientists often fail to recognize the power and indeed agency of material items or environments. Pels, for example, points out that materiality is not simply a quality of objects, rather that materiality resides in the sensuous processes of 'human interaction with things' (1998:100). In his view, 'the "material" is not necessarily on the receiving end of plastic power, a tabula rasa on which signification is conferred by humans: Not only are humans as material as the material they mould, but humans are moulded, through their sensuousness, by the "dead" matter with which they are surrounded' (Pels, 1998:101). And even as early as1981, Csikszentmihalyi and Rochberg-Halton were arguing that 'the impact of the inanimate object ... is much more important than one would infer from its neglect [in psychoanalysis]. Things also tell us who we are, not in words, but by embodying our intentions. In our everyday traffic of existence, we can also learn about ourselves from objects, almost as much as from people' (1981:91).

This chapter therefore not only reincorporates the dead into our discussion of older adults' home space and social identity, but also questions the ontological distinction which is made between human beings and the objects which surround them. It queries the view that objects have only a derivative agency, even though, as Pels points out, 'those people who say that things talk back' risk being seen as 'dangerously out of touch with reality' (1998:94). In defence of the agency of objects, Spyer highlights 'the powers that things have to entrance, raise hopes, generate fears, evoke losses, and delight' (1998:5) – powers that cannot be understood simply in terms of a stable person/object hierarchy.

It remains, nonetheless, to show how these strands of argument help with an investigation of the spatio-temporal environments of later life. We can begin this task by highlighting the tyranny of a linear, diachronic view of time, the notion that friends, family and other loved ones have somehow been left behind, either because their lives have taken a different, independent course – or because they are dead. To make full sense of the materiality of older people's domestic environments, we need a synchronic view of time which can reveal the capacity of material items to mediate time in complex and ambiguous ways. To see life and death as separate and opposed categories, as temporally discontinuous, is to ignore the transcendent valency of memory and material culture. For the people interviewed for this project, material objects positioned within home space had the capacity to animate a former partner's social presence, to make it contemporaneous with everyday life; and moreover to precipitate *new* or varying forms of social engagement. For example, Irene, a widow living in sheltered housing, broke the isolation of her few rooms and forced herself into the outside world in order to replace the cut flowers which stood as her marker of her husband. She was appropriating a task for which he had taken responsibility in life: 'That is my memory of my husband, those flowers in that vase – always', she said, 'And if I, if I haven't been ... well I've made myself go – I think, 'Well, he'll be watching'. Even the constrained space of her flat therefore provided scope for an enduring, routinised practice, if not for the display of many objects from the past. It was the material resource which sustained her former partner's continuing social presence and provided a touchstone in current decisions about going out into public space. Similarly, after Godfrey's wife died he continued to live in the sheltered accommodation they had shared – and continued to go out on the trips that were organised. He said: 'Did the lot [trips]. I didn't think I would be able to do but, oh well, I'll knuckle down. I mean she wouldn't have wanted [me not to]'. Again, these data show a material practice being re-animated in a heterosexual partner's absence.

The data to follow similarly concern the objects or practices of everyday life, the positioning of unremarkable items in sitting rooms, bedrooms, gardens and kitchens. Arguments which suggest that both the dead, and the grief which may be felt for them, have been marginalized or sequestrated ignore evidence of the contemporaneity of the dead and the living, particularly within the domestic spaces of the home. Take, for example, the chair belonging to the deceased partner, its spatial location still making it a focal point for conversation, and, when used by visitors, an embodiment of their continuing presence in the home and participation in interaction. The materiality of everyday objects and places, their endurance across time, gives them a special relationship with the human lives which many of them will survive. Though human beings currently inhabit environments which contain many short-lived manufactured items, for many, a significant proportion of their material surroundings is still inherited from previous generations – and will stand long after the end of their embodied lives. Indeed many manufactured items have a highly problematic longevity, willed on future generations as non-biodegradable rubbish. Thus an investigation of the social life of things uncovers epochs and eras co-existing in the spaces of the present, whether their reach is far into the past or onwards into an indeterminate future. Within the times and spaces of the everyday world, therefore, the dead and their loss intervene as imperceptible reminders of what was, is, and will be.

Once we shift our gaze from the sacred, institutionalised and bounded spaces of 'death ritual', and begin to recognise the *ritualisation* (Seremetakis, 1991) which passes for everyday life, the material contexts of older people's lives take on a different set of meanings. As Unruh argued (1983:345), much of what survivors are left with after a death is 'ordinary and mundane'. What he describes is an important reinterpretation of the mundane, or a sifting through of the deceased's life, 'to preserve – and create – identities' (1983:345).

His/Hers/Ours/Mine

The project's focus on bereavement of a heterosexual partner is rooted in a concern with gender issues. It asked about the extent to which shopping and leisure environments are geared towards the heterosexual couple; whether gender stereotypes govern which member of a couple gets to go out alone and to where; and whether women rather than men control domestic routines and decision-making about household space. However by making heterosexual partnerships our focus, the gendering of the *inanimate* became evident. After a death, the spaces which remain, the objects left behind, and the practices now uncompleted, stand as social trails of particular gendered persons; as Parkin says, 'extend(ing) that personhood beyond the individual's biological body' (1999:304). Given that our personhood is profoundly gendered, therefore, those social trails speak to and of the femininity or masculinity of the deceased. For example, when Ash (1996) describes the men's ties she is left with after her husband's death, she says that while the 'feel' or 'texture' of fabric is not necessarily gendered, the identity of particular garments often is. Among all her husband's clothing, the ties are one of the most peculiarly male items of clothing left. As she says, 'it is precisely through a *man's death* that the existence of the ties left, signifies and underlines the absence of his presence to me as *female, living* or present' (Ash, 1996:220). Working from this perspective therefore, we become sensitised to what Kirkham and Attfield describe as, 'the myriad ways in which the dynamics of gender relations operate through material goods' (1996:1).

Kirkham and Attfield go on to highlight 'the pervasiveness, persistence and power ... of binary oppositions ... which play a part in the gender differentiation of many objects' (1996:4). Thus gender differences are often conceptualised in terms of a hierarchised binary where 'masculinity' and 'femininity' are constructed as oppositional categories (Dunne, 1999). And as Kirkham and Attfield argue, 'in our society today, the main visual oppositions which cluster around that of male/female include light/dark, pink/blue and large/small, although others such as geometric/organic, smooth/rough and hard/soft also apply' (1996:4). With respect to life after the loss of a heterosexual partner, therefore, many of the items of material culture inherited by the surviving individual remain not simply as reminders of that person and their death, but also as now unclassifiable representations of a lost heterosexual relationship. Particularly for older couples whose gendered identities may have followed more traditional or stereotypical paths, the domestic environment of a man cannot easily accommodate a woman's cosmetics and toiletries; they exceed the categories of masculine domestic space. The same might be said for a man's pipe, work tools or sports equipment in what is felt to be a woman's space.

Here, work on the fetish object within material culture studies is helpful (Spyer 1998). The fetish, it is argued, is an object which cannot easily be contained within any one classificatory system. Indeed it not only transgresses the boundaries of established systems but actively spills out across the limits of those spaces which might be allocated to it (Shelton, 1995:7). Thus Pels (1998:99) has described how the fetish 'threatens to overpower its subject, because – unlike our everyday matters – its lack of everyday use and value make its materiality stand out, without much clue as to whether and how it can be controlled'. After a death, those mundane items of everyday usage such as perfume, wallets, shoes and hats take on precisely these qualities. Rendered use-less with the loss of their previous owner, their persistent materiality can obtrude into a present where they cannot easily be incorporated into a new scheme of things, nor can they be thrown away. As Dunne says, 'when men and women form heterosexual partnerships gender *difference* is being affirmed in the everyday routines of social life' (1999:71). So when, for example, flowers as opposed to stripes are used to signal the gendered ownership of items which share precisely the same purpose, such as travel wash bags, we see evidence of what Dunne describes as 'the suppression of similarities between women and men' (1999:71). The same can be said with regard to children and parents who share a social world within which 'childhood' has increasingly taken on the status of a separate social category – one demarcated through a whole range of specialist products, from toys and clothing through to video games and magazines (Hockey and James, 1993). What we focus on here is indeed 'the social life of things'. Residual objects which extend the personhood of someone belonging to a social category marked out through difference – whether a heterosexual partner or a child – can be seen to have a particularly resonant and potentially problematic materiality. With a high symbolic value, and possibly a not insignificant economic worth, they cannot easily be jettisoned *or* used.

Giving clothing to a family member of the same sex is one way of subverting its fetishistic power. The giver directs it into an appropriately masculine or feminine system of apparel. Yet sometimes even then it cannot so easily be detached from the body of its original wearer, a point which bears out Pels' argument that materiality resides in the sensuous processes of 'human interaction with things' and not just in the objects themselves. To repeat, 'Humans are as material as the material they mould' (1998:100/101). In his autobiography, the actor, Anthony Sher, also describes the inseparability of his father from the apparel which survived him. This was born in on him very powerfully when he attended his father's memorial service in clothing left over in his father's wardrobe: 'It was odd, I felt I was climbing into a costume; preparing to play Dad. Later, going for a pee, I experienced an even stranger sensation: reaching for myself through his flies' (2001:63/64).

Data from Nancy, a widow interviewed as part of the project presented here illustrates the *variety* of meanings which the belongings of a heterosexual partner can take on. She would not give her husband's clothing to a charity shop because she felt that items would be sold off cheaply: 'Get five pence', she said, 'he was worth more than that'. So she gave them to the institution where he had been cared for when he died, knowing from experience how often residents needed changing: 'They hadn't no bank of clothing, so that's where they all went … fine woollies … a forty pound pair of shoes, still had the labels on'. Giving them away had upset both Nancy and her son, but, she said, 'I knew his clothes were doing … some good' because he had often come home in somebody else's trousers when he'd been cared for there. His residual good quality clothing therefore took on additional worth by ameliorating the bleakness of the institutional setting, albeit retrospectively. His wallet, by contrast, she described as 'the tattiest old thing you've ever seen'. Unlike her concern with the financial value of the clothing – and by implication the symbolic worth of her husband, the wallet she described as 'worn out'. Yet she kept this used-up item with its mark of daily handling by her husband, integrating it into her scheme of things by classifying it as a memory object: 'I've got that in my special memory box', she said. Neither the clothing nor the wallet therefore took on an excessive or fetishistic status. In choosing particular spatial locations – the institution and the special memory box, Nancy was able to classify or accommodate them symbolically. However, the jigsaws which her husband would complete and glue to hardboard as his winter hobby, represented far more problematic items. They represented many hours of concentrated work on his part, yet Nancy did not judge them suitable for giving away, had difficulty finding space for them, but could not jettison them as 'rubbish'. 'I had them all over the house …and I said "Oh, I don't like this" … but I can't throw them away'. Some stood up in her wardrobes. 'And he's got one in his shed', she said 'and it's slowly disintegrating and I can't take it down'. Pieces of it were melting with the heat of the sun on the shed. 'Every time I go in there there's some more jigsaw's gone', she said. When the last piece fell off she intended to throw it away. The disintegrating jigsaw can be seen as an object with considerable agency in that it is shaping the trajectory of Nancy's process of grief, a gradual decay to which she is responding. In allowing it this agency, Nancy was leaving her husband's long hours of work to undo themselves, so sparing herself this painful task. The jigsaws location within the shed also underscores the gendered nature of these environments of memory. Nancy described it quite explicitly as 'his shed', a form of gender differentiation which reflects the 'polarity of room coding' which became evident from the nineteenth century onwards (see Kinchin, 1996). Not only does she leave his jigsaw there to slowly disintegrate; she cannot appropriate this space of 'his' as her own. She keeps her own gardening tools in the more feminised space of the scullery, items which she cannot, as yet, give space to alongside the disintegrating jigsaw in the shed which so powerfully evokes her husband's absence.

The task of understanding loss as a materially-grounded process of transition therefore involves exploring the gender-specific categorical identities of aspects of widowed individuals' environments; and, in particular, those 'hard to place' items which have a particularly problematic materiality. Kelleher uses the bodily metaphor of 'connective tissue' to describe the clutter which goes to make up the shared materiality of the human being and their physical environment (2001:224). Rochberg-Halton makes a similar point with the statement: 'Household artefacts do not exist atomistically, except perhaps in a museum display; rather, they form part of a gestalt for the people who live with them – a gestalt that both communicates a sense of home and differentiates the types of activities that might be more appropriate for one part of the home than another. Hence, the meaning of things one values are not limited just to the individual object itself, but also include the spatial context in which the object is placed, forming a domain of personal territoriality' (1984:353). Spaces, items or practices can be reinterpreted at death, in a series of potentially gendered readings of the kind alluded to by both Unruh (1983) and Ash (1996). Within heterosexual households, they can be identified and then re-identified as, for example, 'his', 'hers', 'ours' or 'mine'.

The data under discussion here suggest that passage along the living/dying/mourning trajectory can involve all kinds of re-interpretive processes. To give just two examples, that which in the past was 'his' or 'hers' can be appropriated in the present as 'mine'; and that which was once 'ours' can be experienced as painfully 'excessive', an out-of-time/place reminder of a heterosexual relationship which can no longer be lived out – and may therefore be re-cast as 'his'. Thus, Irene, the widow described earlier, who went out each day to buy cut flowers, was making her husband's former routine her own, and in so doing, connecting with him in a way which made this spatialised practice 'hers' – and indeed the memory evoked as a result was 'mine'. By contrast the pictures and plates which had hung on the walls of another couple's shared sheltered home were removed by a widower after his wife's death: 'I took the pictures off the wall because I didn't like them … plates and all sorts of things that she liked'. He described how they had divided the space in the flat to accommodate their different interests, adding 'Now of course I just spread my … it's all mine so I spread it where it goes'. It is interesting to note that although sheltered housing offers little scope for the display or rearrangement of significant objects, the transition to single status can still be marked by *gaps* or *absences*, as formerly shared items become re-cast – and perhaps rejected – as 'hers' or 'his'.

This scope for reassigning gendered identities to objects or spatialised practices after a death provides a new slant on work conducted by Pearce (1998). Her questionnaire survey of randomly selected individuals investigated the meaning and value of objects for people of different genders and social class backgrounds. She found that while men ranked vehicles highest in importance, women put jewellery first, going on to argue that 'for women, chosen objects are a mode of organizing the web of inter-personal relationships which bind together the past, present and future in a tangible, visible network. The objects symbolize shared memory, experience and emotion, and their presence in the home or on the person creates feelings of security' (1998:239). 66 per cent of Pearce's sample were partnered, and, as noted, women focused far more on the emotional, past-oriented nature of objects, with men having a more instrumental set of preferences.

Though generalizations on the basis of our relatively small study must remain speculative, it nonetheless seems to be the case that in the *absence* of a female partner, men were appropriating the emotional work of 'organising the web of inter-personal relationships which bring together the past, present and future' (Pearce, 1998:239). Bill, for example, in his late sixties, widowed for approximately two years, brought up the topic of his wife's jewellery and spoke about it at length. As a gendered item, the jewellery was clearly 'hers'. Yet, in that much of it had been given to her by Bill, he clearly saw it now as 'his'. And in the plans he described for redistributing some of it among the family, it was not just this set of gendered objects, but also the emotional labour of women, which Pierce (1998) describes, that he was appropriating. He said 'There is a lot of memories and I've some pearls, the first Christmas present I ever bought her, I've still got them. And I still have her jewellery box with a lot of jewellery in it, I've her engagement ring … and I had a watch in, it still ticks … and nobody said "Could they have it?" so I thought "Well I'll keep that …"'. But of the gold jewellery (earrings and a necklace) he had bought his wife on a foreign holiday just before her death, he said 'Different members of the family that said "Could they have it?" I said ... "That's mine"... I said, "If I get a granddaughter, that's where they'll go ... if not", I said, "When Martin (his young grandson) gets married, his wife gets 'em". I said "They're not going out of the family"'. His sister-in-law in Australia would have loved it, he said, but: 'I'm not letting it go. I said, I want it where I can see it'. Here we see the indivisibility of the living and the dead, mediated via key material objects. Like the women represented in the study by Pearce (1998), Bill's wife had prized her jewellery for the sentimental occasions associated with it. Worn next to the skin and now contained within the jewellery box, these items embody times and places spread across decades and continents. Rather than letting them lie as inanimate objects, Bill orchestrates a set of spatio-temporal social relationships within which he locates both himself and his wife, as well as grandchildren not yet born, and a future generation of in-laws as yet unknown. The materiality of the jewellery, bought during the period of emotional intensity leading up to his wife's death, is important as a resource through which Bill plans to mark out the extent as well as the limits of 'family'. In addition, it enables an immediate, sensual relationship which Bill intends to sustain, for the present: 'I want it where I can see it'.

Conclusion

This chapter set out to identify those theoretical perspectives which allow us to work within the gaps between literatures which are currently largely silent on the topic of later life bereavement. By drawing on approaches from cultural geography, feminist readings of gender, and debates about agency within the field of material culture studies, it has shown the value of a spatialised account which transcends a whole series of categorical boundaries: between the past and the present, the living and the dead, the animate and the inanimate. One final point to note is that while spaces and the objects they contain are often understood to be distinct from one another, what these data show is their indivisibility. The older women and men we interviewed described objects shaping and – in the event of a death – reshaping the spaces which appear to do no more than contain them. In reading these accounts we learn how the material environment which survives a dead spouse can take on agency and provide a powerful form of emotional mediation which orchestrates the affective life remaining to their bereaved partner.

References

Appadurai, A. (1986) *The Social Life of Things: Commodities in cultural perspective*, Cambridge: Cambridge University Press.

Ash, J. (1996) 'Memory and Objects', in P. Kirkham (ed.) *The Gendered Object*, Manchester: Manchester University Press.

Bennett, K. (1997) 'Widowhood in elderly women: the medium- and long-term effects on mental health', Mortality, 2(2), 137-148.

Bornat, J. (1994) *Reminiscence Reviewed: Perspectives, Evaluations, Achievements*, Buckingham: Open University Press.

Bytheway, B. (1995) *Ageism*, Buckingham: Open University Press.

Csikszentmihalyi, M. and Rochberg-Halton, E. (1981) *The Meaning of Things: Domestic symbols and the self*, Cambridge: Cambridge University Press.

Coleman, P. (1994) 'Reminiscence within the study of ageing: the social significance of story', in J. Bornat (ed.) *Reminiscence Reviewed: Evaluations, Achievements, Perspectives*, Buckingham: Open University Press.

Dunne, G.A. (1999) 'A Passion for "Sameness"? Sexuality and Gender Accountability', in E.B. Silva and C. Smart (eds) *The New Family*? London: Sage.

Fairhurst, E. (1999) 'Fitting a quart into a pint pot: making space for older people in sheltered housing', in T. Chapman and J. Hockey (eds) *Ideal Homes? Social change and domestic life*, London: Routledge.

Field, D. (2000) 'Older people's attitudes towards death in England', Mortality 5(3), 277-298.

Hallam, E. and Hockey, J. (2001) *Death, Memory and Material Culture*, Oxford: Berg.

Hockey, J. and James, A. (1993) *Growing Up and Growing Old. Ageing and Dependency across the Life Course*, London: Sage.

Kellaher, L. (2001) 'Shaping Everyday Life: beyond design', in S.M. Peace and C. Holland (eds) *Inclusive Housing in an Ageing Society: Innovative approaches*, London: Policy Press.

Kinchin, J. (1996) 'Interiors: nineteenth-century essays on the "masculine" and the "feminine" room', in P. Kirkham (ed.) *The Gendered Object*, Manchester: Manchester University Press.

Kirkham, P. and Attfield, J. (1996) 'Introduction', in P. Kirkham (ed) *The Gendered Object*, Manchester: Manchester University Press.

Klass, D., Silverman, P.R. and Nickman, S.L. (1996) (eds) *Continuing Bonds: New understandings of grief*, Washington: Taylor and Francis.

Parkin, D. (1999) 'Mementoes as Transitional Objects', *The Journal of Material Culture*, 4(3), 303-20.

Pearce, S. (1998) 'Objects in the Contemporary Construction of Personal Culture: Perspectives Relating to Gender and Socio-Economic Class', *Museum Management and Curatorship*, 17 (3), 223-241.

Pels, P. (1998) 'The Spirit of Matter: On Fetish, Rarity, Fact and Fancy', in P. Spyer (ed) *Border Fetishisms. Material Objects in Unstable Places*, London: Routledge.

Riches, G. and Dawson, P. (1998) 'Lost Children, Living Memories: the role of photographs in processes of grief and adjustment among bereaved parents', *Death Studies*, 22, 121-140.

Rochberg-Halton, E. (1984) 'Object relations, role models and the cultivation of the Self', *Environment and Behavior*, 16 (3), 335-368.

Seremetakis, C.N. (1991) *The Last Word: Women, death and divination in inner* Mani, Chicago: The University of Chicago Press.

Shelton, A. (1995) (ed) *Fetishism: Visualising power and desire*, London: Lund Humphries Publishers.

Sher, A. (2001) *Beside Myself*, London: Hutchinson.

Spyer, P. (1998) (ed) *Border Fetishisms. Material Objects in Unstable Places*, London: Routledge.

Thane, P. (1983) 'The history of provision for the elderly to 1929', in D. Jerrome (ed) *Ageing in Modern Society*, London: Croom Helm.

Unruh, D. (1983) 'Death and Personal History: strategies of identity preservation', *Social Problems*, 30(3), 340-351.

Chapter 11

'Looking in the Fridge for Feelings':[1] The Gendered Psychodynamics of Consumer Culture

Colleen Heenan

Introduction

> Cause I've sat there and thought, 'I am not going to give in. I am not going to go down to the biscuit tin', so I've made myself a drink and I've sat there and I have visions of the biscuit tin and what's in it and the longer I leave it, the stronger the urge gets. It's as if the biscuit tin's shouting, 'What about me?' ('Lyndsay', a pseudonym for an anorexic participant in a women's eating disorders therapy group).

Food and its consumption create spatial relationships between consumers and the food they eat. These spatial relationships range between the global reach of those associated with food industries and the local scale of individual embodiment (Bell and Valentine 1997). This chapter focuses on the latter, with specific reference to the experiences of women with eating disorders. My approach draws on my experience as a psychodynamic psychotherapist, and I make strategic use of material from a women's eating disorders therapy group to elaborate these points. Conceptually, the analysis I offer draws on feminist psychoanalytic and critical theory which makes clear the closely inter-twined relationship between subjective experience, or the internal and the social world (Butler 1997; Bordo 1988; Orbach 1986).

In the extract above, Lyndsay describes an intense emotional interaction that takes place within her, and between her, the biscuit tin and its contents. Moreover, within the context of the therapy group, Lyndsay also explains that the biscuits in the tin were purchased by her for the men in her family, suggesting that her gendered position within her family and in the wider world are also relevant to understanding the emotional geographies giving rise to the experience she recounts and to the difficulties bringing her to the group. This chapter offers an account of these emotional geographies.

While not every woman develops an eating disorder, continuities between supposedly problematic and non-problematic experiences of eating are well known (Orbach 1978). However, male readers may be shouting 'what about me?' in response to this focus on women's experience. Given that consumer culture has increasingly generated 'disciplines of masculinity' that construct and apparently assuage anxieties about the performance of masculinities in similar ways to those more traditionally associated with femininities, the argument I make in this chapter could probably be extended to men. However, I do not attempt that task.

Bloom, Gitter, Gutwill, Kogel and Zaphiropoulos (1994) maintain that consumer culture is able to bridge the public/private binary in complex ways that have particular consequences for how women are able to act upon and within their environments. They suggest that consumerism exploits beliefs in individual choice alongside hopes of transformation in such a way that it not only taps into but also constructs individual desires, whilst simultaneously stirring up fears of exclusion (also see Cushman 1995). This mirrors psychoanalytic accounts of infant development, in which psychic separation and individuation are understood as inextricably linked to the capacity to relate to others as well as being linked to developmental transformations (Winnicott 1971; Bollas 1987), and unconscious hopes and dreads (Mitchell 1993). Psychodynamically, consumerism can be conceptualized as extending the arena in which unconscious transactions are expressed beyond the boundaries of the family.

Human development in contemporary western society takes place in specific circumstances in which food consumption goes hand in hand with inherently gendered prescriptions about body size and shape that make women especially vulnerable to alienation and exploitation. Consumerism appeals to women's gender-specific sense of agency; that is, the necessity of 'transforming themselves' and 'making things right for others' (Orbach 1986). At the same time, the use of women's bodies as 'floating signifiers' for the diet, food and fashion industries, along with the inbuilt tensions between the 'intense stimulation of and simultaneous frustration of desire' (Gutwill 1994a: 37) inherent within these trades, constructs and maintains a gendered sense of individual emptiness (Cushman 1990).

In this chapter I explore the ensuing emotional geographies that contemporary western women negotiate. I make use of Bloom *et al.*'s (1994) thesis that women are incorporated into consumer culture[2] via its extension of the 'relational matrix' of the family; that is, by means of their attachments to the complex relations through which gender is constituted (cf. Butler 1990). Bloom *et al.* argue that these attachments are tainted because consumerism is 'subject seeking' in ways that are particularly 'toxic' to women. As such, consumerism is able to exploit women's gendered identifications through their participation in the 'disciplinary practices of femininity' (Bordo 1988) while simultaneously reproducing itself.

In order to elaborate this argument, I first summarise aspects of the psychoanalytic view of feeding and development, particularly the British school of 'object relations' theory. I discuss connections between object relations and consumer culture before turning to the case of the diet industry and to its gendered dynamics. In order to further illuminate the emotional geographies, I then return to the opening extract and other clinical material from 'Lyndsay'.

Feeding and Child Development: A Psychoanalytic Perspective

The sensation of hunger 'is one of the most basic and discrete avenues by which need is learned about altogether' (Bloom and Kogel 1994: 41), and it is therefore no surprise that psychoanalytic theory regards food and feeding as having such key functions in child development. Bloom *et al.* (1994) draw on Winnicott's version of British object relations' theory (Greenberg and Mitchell 1983), to suggest that food's material and symbolic connection with the infant's primary caretaker makes it a key 'transitional object' in development and one that continues to play a significant role throughout life. From a psychoanalytic perspective, the feeding process involves not just physical consumption but the unconscious 'introjection' of the relationship between infant and caretaker. Introjection does not simply create a mirror of external relationships. Rather, the infant incorporates into their subjective, interior world, specific emotional qualities of the interpersonal feeding relationship. As such, the infant's psychic life is 'peopled' with unconscious representations of aspects of others, or 'objects', which are then further acted upon internally. Object formation occurs through 'primary process' activity, which condenses thoughts, creating metaphors through symbolization or association. Primary process also 'displaces, separating the affect invested in an idea from its content or several elements of an ideational complex one from the other' (Chodorow 1999: 42).

The psychophysical experiences of hunger, food and feeding play a central part in object formation both materially and symbolically. Psychoanalytic theory regards the infant as lacking an innate sense of differentiation between self and other, suggesting instead that external relationships are initially experienced by the infant as internal (Perlow, 1995). The repetitive need to express hunger and the caretaker's response to this need provide ongoing opportunities for the infant to develop a sense of agency and security (if I cry, I know someone will come) and psychic boundaries (this is me, that is you). The infant's sense of agency is thus inextricably bound up with another's response, confirming its inter-subjective and interdependent basis.

Winnicott (1971) suggests that a key element of the psychological developmental process of separation and individuation involves the infant projecting feelings about itself and others onto 'transitional objects' (teddy bears, blankets and food, for instance). The infant's capacity to creatively imbue these objects with emotional and symbolic meaning enables them to be used to manage feelings, impulses and desires. In turn, these transitional objects can counter any loss of control experienced by the infant as a result of distance from the caregiver, thereby facilitating a growing sense of autonomy. While these developmental achievements are gained in part through improved physical mobility and verbal capacity, the infant's increasing capacity to feed itself is also crucial. In addition, with the development of a sense of its own separateness, the infant begins to experience the emotional pleasure of sensing that a separate other takes part in assuaging needs. Autonomy can be further effected through the infant's expression of a wish for, or a rejection of a specific food; these actions represent attempts to establish a sense of self distinct from others, as well as a more finely-tuned expression of desire.

This account presents an emergent geography of emotions mobilised through feeding. The infant's initial experiences are spatially undifferentiated, and the differentiation associated with growing separateness and autonomy are made possible by psychophysical processes that generate complex subjective geographies. 'Inside' and 'outside' remain deeply and always emotionally interwoven (compare Sibley 1995).

Consumer Culture and Object Relations

To a substantial degree the power of consumer culture stems from its capacity to mobilise the psychodynamics and emotional geographies I have just described. Bloom *et al.* suggest that consumer culture can be considered as a form of '"maternal" matrix to which individuals consciously and unconsciously attach' (1994: xiii). Active participation in, and attachment to, consumer culture resonates with individuals' developing psychological structures, such that public culture functions as 'another facilitating environment for intra-psychic life and for people to feel interpersonally connected' (Gutwill 1994a: 18). Consumerism is able to bridge the gap between the public and the private through the ways in which dominant cultural symbols are embodied in consumer objects that actively encourage attachment through possession (Cushman 1995). As such, the role of the primary caretaker is that of the 'female culture mother' (Bloom and Kogel 1994: 49) who directly and indirectly imparts the knowledge required to enact subjectivity – directly in the feeding process through the use of manufactured baby foods, and indirectly through the use of television as 'baby-sitters' (Gutwill 1994a: 22). While immediate caretakers are busy elsewhere, consumer culture actively steps in to function as a cultural parent. The emotional geographies initially generated between infant and caretakers are thus extended beyond the boundaries of family life via consumer culture.

This 'subject seeking' nature of consumerism contrasts sharply with Winnicott's (1971) notion of culture (and mothering) as a benign and private transitional space in which 'the cultural symbol *can become* a channel for relationship' (Gutwill 1994a: 20, emphasis added). Indeed, the emotional and economic power of consumerism is that, like the infant's experience of primary caretaker, its cultural symbols seem to 'know' not just what the consumer wants, but what she or he *needs*. Moreover, it is the culturally symbolic meanings attached to such things as children's toys associated with popular films, which accounts for the way mass culture is actively introjected and generates attachments. Transitional objects are also imbued with culturally specific notions of masculinity or femininity. For instance, adolescent attachment to culture comes through the personal reproduction of culturally determined 'ways of being'. This functions as an adolescent 'rite of passage' through which 'looking good' is equated with 'being right'. Looking 'good' means having clothes with the 'right' label. For girls, looking good and being right are equated with having the right body size and shape (Bloom *et al.* 1994).

Further, consumerism positions the consumer as agentic; someone who is free and informed enough to make choices about consumption, thus encouraging active participation (Cushman 1995). However, while it stirs up a desire for more – for example for 'better' goods – it simultaneously stirs up fears about exclusion and worth. At the same time, it is essential that consumers come to regard themselves as exercising their democratic right to choose freely, rather than see themselves or the producers of goods, as greedy.

The insidiousness of consumer culture is that, by encouraging consumers to look after themselves by 'getting the best' at the same time as prescribing what is 'best', it appears to provide the very combination of nurturance and authority associated with parental roles. It can also appear to anticipate needs by generating desire while simultaneously providing goods as if they were required. Paradoxically, in order for consumerism to reproduce itself, its products must have an inbuilt lack of durability or failure to satisfy, so that more goods continue to be required. Hence, one of the more 'toxic' aspects of consumerism is its capacity to locate blame for failure and disappointment within the individual and not itself. A dynamic emotional geography is thus invoked, in which feelings move between, and confuse distinctions between, interior landscapes of object relations and consumption landscapes replete with objects to which aspects of emotional life attach.

The powerful psychodynamics of the continual 'seduction and rejection' of consumerism can be illuminated through Fairbairn's (1952) discussion of object relations theory. Fairbairn suggests that the infant's dependency on the (m)other requires it to blame itself when caretaking goes wrong, a psychological defense mechanism that Fairbairn called the 'moral defense'. This enables the infant to retain some kind of attachment to the (m)other in the face of disappointment. Further, the infant retains a sense of omnipotent control by internalizing only 'bad' objects; that is, those linked with unhappy feelings. While this preserves (m)other as 'good', it results in an *internal* sense of persecution, which Fairbairn described as an 'internal saboteur'. Blaming the self and not the other enables the infant to turn itself into whatever the internalized bad object (the 'persecutory superego') suggests (symbolically) is required. This defense mechanism has two functions: it not only safeguards the possibility of *future* assuagement through self-transformation; it also preserves the other (and self) from the disturbing feelings that have arisen. Moreover, the dynamic nature of unconscious processes mean that internalized representations continue to be acted upon, and themselves act upon other aspects of the ego structure, throughout life such that the defense continues to be employed in a more generalized way (see Greenberg and Mitchell 1983 for a fuller account).

Dieting, Thinness and Gendered Embodiment in Consumer Culture

Bloom *et al.* (1994) maintain that the diet industry represents one of the most gendered and tenacious ways in which consumerism 'seeks out' women. It is tenacious because it both tantalizes and seduces through the symbolic happiness that the diet represents. At the same time, it threatens isolation, rejection and punishment should the consumer not respond. Of course, dieting is 'a business that thrives on failure' (Gutwill 1994b: 32), despite failure being deemed to be the fault of the woman rather than the diet. This dynamic functions to pathologize individual women while simultaneously providing a means to redemption – through another diet. However, in order to understand why dieting has become such a key part of consumer culture, it is necessary to explicate the centrality of the gendered body in contemporary western society.

Featherstone (1982: 170) describes consumer culture as 'latch[ing] onto the prevalent self-preservationist conception of the body, which encourages the individual to adopt instrumental strategies to combat deterioration and decay (applauded too by state bureaucracies who seek to reduce health costs by educating the public against bodily neglect) and combines it with the notion that the body is a vehicle of pleasure and self-expression'. The dual emphasis on maintenance and appearance not only provides endless possibilities for the production and consumption of commodities but also allows for their conflation within the social practices of, for instance, shopping, dining out and exercising. Further influenced by the growth of 'narcissism' (Lasch 1979), life in contemporary western society has become a 'project of the self' in which the 'performing subject' cultivates inner and outer self (Giddens 1991). The private concerns of the inner body, expressed as both physical and mental health, become a vehicle for not only appearing but also for participating in public life. At the same time, appearance and participation are subject to specific prescriptions.

The cultural importance and gendered significance of thinness in contemporary western society can only be understood when it is juxtaposed with its hatred and dread of fatness. By the mid-twentieth century, thinness in North American society came to be associated with both independence and upward mobility while also representing a shift away from the poverty and obesity associated with an 'immigrant' status (Gutwill 1994b). Bordo (1990) argues that hatred of fat is so prevalent because it challenges the equation of 'discipline' with 'happiness'. As such, thinness equates with moral constraint or self-discipline. The slim and 'toned' body is the epitome of the managed and useful body within consumer culture. Fat comes to signify the 'body's potential for excess and chaos' (Bordo 1990: 89).

In addition to performing and cultivating the self, the subject also regulates the self. Of relevance here is Foucault's (1978) theory of 'bio-power' in which self-discipline replaces overt dominating power (Fairclough 1992:50). Foucault argued that the modern subject has been morally constituted through, for instance, the generation of knowledge about bodies and the mind by the medical and social sciences. Power derives from knowledge that incorporates the subject, producing and transforming subjectivities. The modern subject is not only 'productive' but is also apparently free to exercise choice/knowledge through the disciplines of bio-power such as the pedagogical use of examination, including self-examination (Rose 1990; Fairclough 1992). However, there is a further paradox of consumerism: while encouraging consumption, it also promotes regulation, particularly in relation to the body. This constructs a perpetual dilemma: how does one take up a place in culture whilst simultaneously exercising moral and physical constraint?

Foucault's (1978) notion of *discursive* 'panopticons' offers a way of understanding the gendered regulation of the body in consumer culture. The term 'panopticon' refers to the way in which prisons were designed to optimize scrutiny. Bartky (1988) and Bordo (1988) suggest that through the private and public practices of discipline such as femininity, women are not only constantly observed but also learn to observe themselves and others without apparent coercion (also see Smith 1988). As Bartky further elaborates: 'In contemporary patriarchal culture, a panoptical male connoisseur resides within the consciousness of most women: they stand perpetually before his gaze and under his judgment' (1988: 72). The insidiousness of modern discipline is that it provides the means for a sense of accomplishment, of being in control, of identity: the experience of subjectivity is dependent on not just what one knows, but 'knowing what to do' (Bartky 1988: 77).

Gutwill (1994c) also proposes that the promotion of thinness as equated with beauty, is a subtle expression of women's contradictory social position: they can 'have it all' but at a price, and the price is to take up less space, and to remain within a well-defined space in the modern world. For women, thinness has been promoted as the ideal body image for the public workplace, marking a separation from the 'roundness' of a motherly (private, domestic) figure. Thinness may also function as a way of 'de-sexualizing' women's bodies and thus 'removing temptation' or 'feminine frivolity' from the seriousness of the productive domain. MacSween (1993) points out that the ever-changing corporeality of women's bodies through menstruation and pregnancy, presents an antithesis to the rational desire of modernity, to 'know everything', of 'mind over matter'. Orbach (1985:89) regards this gendered signification as a 'mass internalized misogyny' wherein infantile anxieties about being able to control the mother through her body become externalized and projected onto women's bodies wholesale, as something to be feared and controlled. This may be one reason that (specific) women's bodies have increasingly become attached to the 'floating signifiers' (Featherstone 1982) of consumerism.

This use of the female body as a gendered consumer signifier breaches the public/private split in complex ways. It offers women the means to attach to the world outside the home through setting up a 'hall of mirrors' in which women constantly see other women (or is it themselves?) mirrored back to them. These 'ideal' (and changing) mirrors, while seeming to reflect themselves as they are now, hold out endless possibilities (choices?) of who they might become – or who they *should* be. However, through constantly viewing and transforming their bodies, women may not feel they 'own' them. As the mirrors of consumerism are both authoritative and nurturing, they can appear to be benign Winnicottian transitional spaces, promising containment and fulfillment of desire (Bloom *et al.* 1994). Yet consumer mirrors de-center the viewer, *constructing* a sense of subjectivity through a false belief in agency. While appearing to offer the means to make fantasy reality, they also manipulate fantasy at an unconscious level, thereby ensuring that *desire* is socially constructed. Confusion about the location of feelings is therefore fundamental to consumerism. Its 'hall of mirrors' encourages women to misrecognise the source and the direction of their embodied feelings.

Eichenbaum and Orbach (1982) have proposed that the ongoing social construction and reconstruction of 'choices' about body size and shape feed into and exacerbate women's underlying sense of body insecurity that derives from the gendered mandate to 'transform the self'. From a psychoanalytic perspective, part of consolidating gender identity involves an early stage in which children seem to want to display and use their bodies, as if to have their sense of corporealism not just confirmed but admired. However, rather than this being a specific developmental phase for females, 'a little girl's mandate to appear (rather than to act or be) and to focus on her appearance is confirmed as intrinsic to her being and equal to being an adequate female' (Bloom and Kogel 1994b: 49). Moreover, 'to some degree, for all women, the critical work of separation, differentiation, and integrating sexuality are displaced on to a struggle to manage one's appetite for food and to transform one's body' (Bloom and Kogel 1994b: 53). Thus, the psychodynamics of gendered subjectivity for women fits in with the consuming and regulatory dynamics of consumerism; both publicly and privately, women 'work on themselves', fantasizing about how they would look *if* they lost weight, or exercised more. The conflicts that arise through participating in rather than resisting the 'disciplines of femininity' can be managed by employing the psychological defense mechanism of 'identifying with the aggressor'.

The insidious and tenacious nature of dieting arises because it fulfills the function within consumerism of offering the means to take up a gendered position within culture – to not respond threatens exclusion. Thus it mirrors Fairbairn's (1952) view of individuals' object worlds, filled with self-blame as a way of managing disappointment with the environment. The symbolic happiness that the diet represents is sought out. At the same time, the underlying threatening aspects of the failure to diet acts as a punishment. Gutwill (1994: 31) describes the resulting internal conversation as: 'Go on – try the diet. If ... if only ... I were good enough, giving enough, sexy, pleasing, or thin enough ... If only I could stay on this diet, I could be acceptable and lovable ... But the truth is that I am *not* good enough; I am selfish, fat, stupid for wanting and needing, ugly, and weak. I deserve all I get. It's my own fault'.

As the designated caretakers, mothers are increasingly subject to intense pressures to curb daughters' physical and psychological appetites (Eichenbaum and Orbach 1982). The continual curbing of women's needs and body size, alongside the encouragement to attend to others, can lead women to construct numerous internal 'false' boundaries, which keep shameful, uncomfortable, forbidden aspects of self separate from each other. This psychological splitting can extend to constructing a false boundary between emotions and bodily experiences, such as hunger, or even between different parts of the body (see Dana 1987; Orbach 1994). Orbach (1978) suggested that women may feel anxious and frightened when their emotional needs are stirred up, unsure not only what the feelings are but additionally anxious because they are not supposed to have either an emotional or physical appetite.

The rapid and simultaneous translation of need into 'hunger' and 'greed' can be understood when it is located within the gendered prescriptions of a 'subject seeking' culture that is built on the paradox of gratification rather than satisfaction.

This 'false feed' (Bloom and Kogel 1994: 42) functions to alienate the consumer from his or her need; the video 'Looking in the Fridge for Feelings' illustrates how, unsure of what she feels or wants, a woman opens the refrigerator door and looks inside, hoping that an item of food might offer her a hint as to what her mood means so that she can connect to it. Eating may gratify the need for some sense of soothing through taking in particular foods that symbolize comfort, or through further disassociation from feelings. Indeed, food is also a 'subject seeking' consumer object and the food industry is able to negotiate a variety of moral positions for itself by promoting 'healthy' food as the 'best option', whilst 'recognizing' that it is not possible to maintain constant constraint.

Of course, the food industry is also able to provide the 'right' food for the 'right' mood – 'junk' food may enable female consumers to 'indulge' themselves, as they so rightly deserve to do, when they work so hard at 'being good' (compare Germann Molz, this volume). But they are then liable to admonish and constrain themselves, labelling themselves as 'greedy', then fat, then ugly, and then in need of losing weight. Undertaking to eat less, they are put in touch very acutely with *physical* hunger, on top of the *emotional* hunger they already experience from being positioned as 'emotional nurturers' to others. By removing food, they may be removing one of the few things which does offer them some comfort and satisfaction, something which makes no demands upon them, something which, when they are left alone with the children and the housework, they can have *all for themselves*.

'What about me?': Feminist Psychotherapy and Eating Disorders

In this section, I use material from a feminist psychodynamic eating disorders group in order to illustrate the emotional geographies described in the preceding sections. In the extract introducing this chapter, Lyndsay describes her ongoing daily struggle with regulating her 'appetite', her hunger and the foods she will allow herself to eat. The seduction of the biscuit tin shouting out, 'What about me?' is the way in which its 'subject seeking' call comes in the guise of an apparently *benign* transitional object that just happens to be at hand in order to meet one's need or desire. Indeed, the blurring of need and desire is an inherent part of its power such that the question 'What about me?' seems to arise from within Lyndsay rather than being generated commercially by the biscuits demanding to be consumed.

As Lyndsay's therapist my clinical understanding is that when she hears a biscuit tin shouting 'What about me?', she is responding to her unconscious split off needs which she has projected onto the biscuits. As a feminist I also understand that, like the biscuits, Lyndsay must keep her needs encased – perhaps within a tin? – in order to 'keep fresh', to respond to the endless demands of her heterosexual subjectivity. She experiences both her emotional and physical needs as greed and thus to be denied assuagement. At the same time, it is rare for Lyndsay to sit down at all, as she is racked with guilt whenever she attends to her own rather than others' needs.

Within a feminist psychoanalytic framework, it can also be argued that Lyndsay is using the label 'fat' to transform her dissatisfaction with her life into a gross distortion of her emaciated body size. In turn, she uses the euphemism of 'dieting' to describe her ongoing self-starvation. However, Lyndsay's gendered sense of persecution not only arises from within but also comes in the form of the 'subject seeking' biscuits, shouting 'What about me?'. Whose 'urge' is she experiencing? Which 'me' is calling out to her? It is not clear how to label, or where to locate her feelings. Is it hunger, located within? Is it temptation, located in the biscuit tin? Or is it a stirring up of emotion, experienced as desire/hunger/temptation? Lyndsay meets the feeling with a gendered act of moral restraint. Despite the fact that she has been a good housewife and mother, she cannot enjoy the tainted reward of the biscuits, which she purchased for the men in her family who also shout (seemingly continually) 'What about me?'. While Lyndsay's emaciated body shouts out for attention, she does not.

In another extract from the group, Lyndsay articulates her gendered and embodied experience of the tenaciousness and insidiousness of dieting. In addition, she also describes the 'transforming' experience of taking in culturally imbued food.

> Lyndsay: … and he [her husband] can't understand why I feel like, like that. I said, 'I feel like Michelin Man'. He says, 'You are now in a size 10 – you've gone from a 14 to a 10. How can you feel like that?' I said, 'Because I do'. I'm like you [Laura], as soon as I eat anything that I know I shouldn't eat, I …
> Laura: You put about two stone on immediately …
> Lyndsay: … feel guilt. I feel guilty if I don't eat it cause I'm thinking, 'Well, I'll have to have something to eat, I'll have to have something'.

When Lyndsay queries her husband's ability to understand the discrepancy between her small body size and her subjective experience of being an outsized 'Michelin Man',[3] she asserts her dubious right to a specifically gendered subjectivity wherein feelings of self-esteem derive proportionally from dress size. However, while dieting has transformed Lyndsay's body, it has not only failed to transform her corporeal sense of self, it has had the opposite effect, such that she feels even bigger than before. Indeed, Lyndsay feels more like a cartoon character than a human. Moreover, she locates the difficulty as arising from within (Fairbairn's 'moral defense'), not in the 'changing hemlines' of consumerism (Orbach 1986), such that size 10 is no longer small enough to induce a gendered sense of well being as a woman in contemporary western society.

This extract also illustrates the culturally imbued nature of consumer foodstuffs such that if Lyndsay 'eats anything she knows she shouldn't', she feels guilty. In turn, the power of moral imperatives associated with particular foods is such that Laura 'puts about two stone on *immediately*', if she eats something she 'shouldn't'. To transgress prescribed notions of a suitable diet incurs immediate punishment in the form of an increase in body weight. The self-blame that ensues is the result of attempting to manage the dilemma of consumerism – to consume but to regulate intake in a 'responsible' manner. Again, it is not clear where to locate the feelings associated with transgression – does guilt arise from within the women or is it situated in the calories of the food? Lyndsay's moral and corporeal dilemma is further complicated by her struggle to be a good 'patient' who is meant to be recovering from anorexia; the necessity for her to eat 'something' belies not just her state of emaciation but also the quandary of contemporary western women who can no longer eat to live but must eat in a regulated manner. As Orbach argues (1986) anorexia speaks volumes about the *contradictory* practices of femininity such that the anorexic is simultaneously highly regulated and pathologized; the 'problem' is that of the individual, not the social.

Conclusion

In this chapter I have drawn on feminist psychoanalytic and critical theory to explore the emotional and geographical dynamics of consumer culture with specific reference to the centrality of the gendered body in contemporary western society. Bloom *et al.*'s (1994) thesis that consumerism strategically transgresses the boundaries between the private and the public in order to produce attachment to a 'parent culture', offers a means to understand how 'subject seeking' consumer culture constructs and exacerbates women's gendered, corporeal sense of agency through promising self-transformation through dieting. The authors posit the emotional dynamics of food regulation as arising from the role of eating and feeding as inherent parts of the process of physical and emotional development. The inherent contradictions between the psychodynamics of choice, transformation and fear of exclusion, make consumer culture particularly 'toxic' for women by providing a culturally sanctioned forum in which women's sense of agency is continually directed towards transforming themselves in a way that literally 'eats at' their psychophysical boundaries. As such, a woman suffering from an eating disorder has ready access to language and cultural goods that facilitate the translation of uncertainty, fears, and need to exercise control into bodily concerns. Not only does she have an 'internal saboteur' (Bloom, Kogel and Zaphiropoulos 1994: 113), who berates her for being a failure, but she lives in a society that has its own socially structured misogynist defense mechanisms (Menzies 1960) to deal with its fear of women's bodies. But it *is* possible to resist prescribed subjectivities or 'consuming' material products. Indeed, as I noted at the beginning of the chapter, not every woman – nor every man – develops an eating disorder. After all, there are limits to the argument that eating disorders are socially produced (Gordon 1990).

While my focus has been gender-specific with reference to women's subjectivities, the analysis offered could be augmented by considering the impact of changing constructions of masculinity. Further, while I have made use of aspects of psychoanalytic theory to illuminate the 'subject seeking' nature of consumer culture, the role of psychoanalysis within consumer culture cannot be regarded as benign. Cushman (1995) makes clear the ways in which consumerism is heavily indebted to psychoanalytic theory's exposition of the dynamics of unconscious processes – theory which it has strategically used in order to obfuscate the location of emotions. In turn, psychoanalysis has both facilitated the growth of contemporary western 'individualistic' culture as well as profited from this through its promotion of psychotherapy as a means to 'cure the ills' of the 'empty self' that inevitably results from living within consumerism (Cushman 1990). As such, the practice of psychotherapy is also 'subject seeking', constructing particular desires, locating emotions, offering salvation through transformation (Rose 1990). While feminist psychoanalytic theory offers a gendered analysis of women's development, there are inherent tensions that arise from its subscription to a body of knowledge, a language of pathology and set of clinical practices located within the emotional geographies of psychoanalysis' modernist notions of the 'self' (see Heenan 1996, 1998).

Notes

1 'Looking in the Fridge for Feelings' (National Film School 1980), is the title of a British video describing feminist therapeutic work with women who eat compulsively.
2 I take the term 'consumer culture' to refer to the shift in late capitalism towards the encouragement of mass consumption by means of advertising that generated discourses of hedonistic lifestyles based on the promotion and gratification of individual needs and desires through the acquisition, use and maintenance of consumer goods. At the same time, the separation of use from value means 'that any particular quality or meaning can become attached to any culture product' (Featherstone 1982: 174). Moreover, the emphasis on visual imagery makes central the display of the body.
3 'Michelin Man' refers to a British television advertisement for automobile tyres featuring a cartoon character whose body is constructed from tyres.

References

Bartky, S. Lee (1988) Foucault, Femininity, and the Modernization of Patriarchal Power, in I. Diamond and L. Quinby (eds) *Feminism and Foucault: Reflections on Resistance.* Boston: Northeastern University Press.
Bateman, A. and Holmes, J. (1995) *Introduction to Psychoanalysis: Contemporary theory and practice.* London: Routledge.
Bell, D. and Valentine, G. (1997) *Consuming Geographies: We are where we eat* Routledge, London.
Bloom, C. and Kogel, L. (1994) Tracing Development: The Feeding Experience and the Body, in C. Bloom, A. Gitter, S. Gutwill, L. Kogel and L. Zaphiropoulos *Eating Problems: A Feminist Psychoanalytic Treatment Model* New York: Basic Books.
Bloom, C., Gitter, A., Gutwill, S., Kogel, L. and Zaphiropoulos, L. (1994) *Eating Problems: A Feminist Psychoanalytic Treatment Model* New York: Basic Books.
Bloom, C., Kogel, L. and Zaphiropoulos, L. (1994) Working Toward Body/Self Integration, in C. Bloom, A. Gitter, S. Gutwill, L. Kogel and L. Zaphiropoulos *Eating Problems: A Feminist Psychoanalytic Treatment Model* New York: Basic Books.
Bollas, C. (1987) *The Shadow of the Object* Free Association Books: London.
Bordo, S. (1988) Anorexia Nervosa: Psychopathology as the Crystallization of Culture, in I. Diamond and L. Quinby (eds) *Feminism and Foucault: Reflections on Resistance.* Boston: Northeastern University Press.

Bordo, S. (1990) 'Material Girl': The Effacements of Postmodern Culture, in *Michigan Quarterly Review. Special Issue: The Female Body*. 24(4): 653-677.

Butler, J. (1990) *Gender Trouble* Routledge: London.

Butler, J. (1997) *The Psychic Life of Power* Stanford University Press: Stanford, California.

Chodorow, N. (1999) *The Reproduction of Mothering: Psychoanalysis and the Sociology of Gender (2nd edition)* Berkeley, Calif: University of California Press.

Cushman, P. (1990) 'Why the Self is Empty', *American Psychologist* 45(5): 599-611.

Cushman, P. (1995) *Constructing the Self, Constructing America: A Cultural History of Psychotherapy* www.dacapopress.com: Da Capo Press.

Dana, M. (1987) Boundaries: One-Way Mirror to the Self, in M. Lawrence (ed.) *Fed Up and Hungry: Women, Oppression and Food* London: Women's Press.

Eichenbaum, L. and Orbach, S. (1982) *Outside In, Inside Out: Women's Psychology, A Feminist Psychoanalytic Approach* Harmondsworth, Middlesex: Pelican.

Fairbairn, W.R.D. (1952) *Psychoanalytic Studies of the Personality* Routledge: London.

Fairclough, N. (1992) *Discourse and Social Change* Cambridge, UK: Polity Press.

Featherstone, M. (1982) The Body in Consumer Culture, in M. Featherstone, M. Hepworth and Bryan S. Turner (eds) (1991) *The Body: Social Process and Cultural Theory*. London: Sage.

Foucault, M. (1978) *Discipline and Punish: The Birth of the Prison*, trans. Alan Sheridan. New York: Vintage Books.

Frosh, S. (1987) *The Politics of Psychoanalysis: An Introduction to Freudian and Post-Freudian Theory* London: Macmillan.

Giddens, A. (1991) *Modernity and Self-Identity* Polity, Cambridge.

Gordon, R.A. (1990) *Anorexia and Bulimia: Anatomy of a Social Epidemic* Cambridge, Mass: Basil Blackwell.

Greenberg, J.R. and Mitchell, S.A. (1983) *Object Relations in Psychoanalytic Theory*. Cambridge, Massachusetts: Harvard University Press.

Gutwill, L. (1994a) Women's Eating Problems: Social Context and the Internalization of Culture in C. Bloom, A. Gitter, S. Gutwill, L. Kogel and L. Zaphiropoulos *Eating Problems: A Feminist Psychoanalytic Treatment Model* New York: Basic Books.

Gutwill, L. (1994b) The Diet: Personal Experience, Social Condition, and Industrial Empire in C. Bloom, A. Gitter, S. Gutwill, L. Kogel and L. Zaphiropoulos *Eating Problems: A Feminist Psychoanalytic Treatment Model* New York: Basic Books.

Heenan, M.C. (1998) Feminist Object Relations Theory and Therapy, in I.B. Seu and M.C. Heenan (eds) *Feminism and Psychotherapy: Reflections on Contemporary Theories and Practices* London: Sage.

Heenan, C. (1996) Women, Food and Fat – Too Many Cooks in the Kitchen?, in E. Burman, G.Aitken, P. Alldred, R. Allwood, T. Billington, B. Goldberg, A.J. Gordo Lopez, C. Heenan, D. Marks and S. Warner *Challenging Women: Psychology's Exclusions, Feminist Possibilities* Milton Keynes: Open University Press.

Lasch, C. (1979) *The Culture of Narcissism* New York: Norton.

MacSween, M. (1993) *Anorexic Bodies* London: Routledge.

Menzies, I.E.P. (1960) A Case-Study in the Functioning of Social Systems as a Defense against Anxiety, in *Human Relations* 19: 95-121.

Mitchell, S. (1993) *Hope and Dread in Psychoanalysis* Basic Books, New York.

Orbach, S. (1978) *Fat is a Feminist Issue* London: Paddington Press.

Orbach, S. (1985) Accepting the Symptom: a Feminist Psychoanalytic Treatment of Anorexia Nervosa, in D.M. Garner and P.E. Garfinkel (eds) *Handbook of Psychotherapy for Anorexia and Bulimia* New York: Guildford Press.

Orbach, S. (1986) *Hungerstrike* London: Faber and Faber.

Orbach, S. (1994) Working with the False Body, in A. Erskine and D. Judd (eds) *The Imaginative Body: Psychodynamic Therapy in Health Care* London: Whurr Publishers Ltd.

Perlow, M. (1995) *Understanding Mental Objects*. London: Routledge.

Rose, N. (1990) *Governing the Soul* London: Routledge.

Sibley, D (1995) *Geographies of Exclusion* Routledge, London.

Smith, D.E. (1988) Femininity as Discourse, in L.G. Roman and L.K. Christian-Smith (eds), with E. Ellsworth *Becoming Feminine: The Politics of Popular Culture* London: Falmer Press.

Winnicott, D.W. (1971) *Playing and Reality* Routledge: London.

Chapter 12

Affecting Touch: Towards a 'Felt' Phenomenology of Therapeutic Touch

Mark Paterson

Prelude

> My first Reiki massage, ever. Anxious because of deadlines, a hundred things whizzing through the head. Talking with the Reiki master, trying to get a sense of what will happen. Then the massage begins, a curious mixture of touch and non-touch. As this is going on, something strange and unexpected starts to occur, then starts to surge uncontrollably, a welling-up that becomes an outpouring. Along with a feeling of incredible release, I start to cry and find I cannot stop.

Introduction: Touching Ambiguities

> We *feel* meanings. A term that indicates the intimate association between bodily senses and emotion. (Game and Metcalfe, 1996:58)

While independence, bodily integrity and self-sufficiency are encouraged in our culture, we also value a more personal, intimate, emotional care in which touch is crucial yet sharply spatially differentiated. It is appropriate in some spatial contexts and with some body parts, but decidedly inappropriate in others. The art of touching is intimately linked with 'body work', ways in which the body is cared for and attended to in settings like medical institutions or therapeutic practices, relating spaces, emotions and power. Something of the dynamic, indeterminate, uncertain, spatial dimensions of touch have been discussed in nursing literature and nursing anthropology, and are relevant to a consideration of a range of therapeutic practices involving touch such as Swedish massage, Reiki, Shiatsu, acupressure and acupuncture. In social theory and research, the body and embodiment has been a site of extensive discursive analysis yet the body's exterior surface, a primary mode of sensing and being in the world, has perhaps received less attention that it deserves. So in part this chapter is a corrective, to reassert touch and tactility in cultures of bodywork, in the relations of carer or therapist and client, and to link this with relevant theory. This will proceed through a theoretical account of therapeutic touch, illustrated with reference to a phenomenological, literally 'felt' experience of Reiki, and with reference to practitioners' understandings of Reiki, their felt connections between bodies, mind and world.

In their editorial 'Emotional Geographies', Anderson and Smith write of the usefulness of 'affect' in geography coming from psychoanalysis (2001:9). However, I will trace 'affect' along an alternative pathway, one that explicitly ties in touch, the body, emotion and space. In nursing and medical research touch is often divided into two types. There is 'professional', 'procedural' or 'task-oriented' touch, which is used in the everyday actual handling of patients. Then there is what Dongen and Elema refer to as 'expressive' touch (2001:154) – a touch that is 'affective' or 'caring' according to Edwards (1998:810). This form of affective touch, touch as expression within therapeutic spaces, is what I wish to explore in this chapter. One of the keys to thinking expressive or affective touch is that, like other therapeutic and psychotherapeutic practices, empathy is invoked (e.g. see Bondi 2003). In this way, feeling is often feeling-with, involving another tactile body, wherein the tactile and the emotional arise within each other. Such feeling-with, the literal touching involved in massage, and the empathic content arising from encounter with another, has interesting spatial effects. For Wyschogrod, empathy and sympathy is a 'bringing near', a drawing of other people into proximity (Wyshcogrod, 1981), and Edwards extends this definition through tactility: 'Touch can be a way of transferring sympathy and empathy between individuals[,] changing the proximity of feeling into what is felt' (1998:810). By its nature then, therapeutic touch is expressive, can open up a non-verbal communicative pathway or channel between bodies, bringing them into proximity, and this can be a deeply affective phenomenon.

Yet tactility in our culture is radically ambiguous, decisively cleaved into acceptable and unacceptable, appropriate and inappropriate (see Thien, this volume). Being touched inappropriately can involve anxiety, fear, disgust or transgression, and these forms of touching are virtually instantaneous in producing such affects. But there are a number of voluntary tactile experiences that explore more acceptable physical-emotional, psycho-social engagements between bodies, between people. In this chapter I discuss the therapeutic qualities of tactile experience. Such sensory experience is emotive, active and passive, shared in and between bodies; *affective*. By examining the therapeutic practice of Reiki I will show the affective power of touch, its empathic and transformative capacity. The therapeutic touch of *Reiki* massage will be used to show the effects of touch on the interactions of body and mind, through a reading of bodily encounter, and the felt experience both on and off the skin surface. I will use Reiki and other touch therapies to talk about the 'deeper' aspects of touch, that is, the ability of touch to communicate inter-subjectively and non-verbally, to open up other therapeutic aspects of experience within prescribed spaces. Such therapeutic experiences are at the edge of representation and expression, from which new ways of thinking about touch, feeling and moving can emerge. I will use *affect* as a way to integrate the somatic and sensory experiences of these tactile, therapeutic practices, and to write of these qualitative, experiential encounters. Like other therapies, Reiki occurs within prescribed yet unstructured spaces, where no formal medical or physiotherapeutic training is necessary. Before examining the case of Reiki in detail, I wish to ground the discussion by exploring the links between *touch* and *affect*, that is, between tactile experience and the affective spaces within which they occur.

Affect

Having introduced the idea of the ambiguity of touching and being touched, one reason to employ the concept of 'affect' is that it is similarly ambivalent, both active and passive in operation, entailing the diametricality of affecting and being affected. One way of thinking affect is through Spinoza, since both Gilles Deleuze (1988) and Moira Gatens (1996) have looked to his philosophy, albeit in different ways, in order to rethink mind and body. Deleuze enthuses about the ability for the Spinozist body to be a series of intensities, to be constantly in relation and connection with other bodies, thereby opening up a discussion of non-differentiated energetics between, and within, bodies. It is an articulation of affect that will be useful in thinking about touch and felt energy between bodies in the therapeutic encounter. Gatens seizes instead upon a different but equally important element from Spinoza, the monism of the Spinozist body, which allows the resolution of mind-body dualism and emphasises the feminist reclamation of the situated, concrete particularity of the body. This section will help clarify the concept of affect and reconcile these disparate trajectories.

Gatens has talked about the female body as intrinsically anarchic or disordered, seemingly at a disadvantage compared to the male body. Yet she finds an unlikely ally in Spinoza, for unlike the notorious mind-body dualism of Descartes:

> For Spinoza the body is not part of passive nature ruled over by an active mind but rather the body is the ground of human action. The mind is constituted by the affirmation of the actual existence of the body, and reason is active and embodied precisely because it is the affirmation of a *particular* bodily existence. (1996:57, original emphasis)

Gatens is able to recover a sense of the body from Spinoza not only as particular, but also as productive and dynamic, that is, to stress 'morphology' over 'biology' (1996:58). This allows the conceptualisation of an alternative *topos* or site to reject the usual dualisms of mind/body, nature/culture and so on. The body's particularity is therefore not limited, but mutable. It cannot be definitively captured, known or limited. If this *topos* is fleshy, mutable, transformable through palpable sensation, then the forces unleashed indicate that the body is dynamic. Gatens continues:

> The Spinozist account of the body is of a productive and creative body which cannot definitely be 'known' since it is not identical with itself across time. The body does not have a truth or a 'true' nature since it is a process and its meaning and its capacities will vary according to its context. *We do not know the limits of this body or the powers that it is capable of attaining.* These limits and capacities can only be revealed by the ongoing interactions of the body and its environment. (1996:57, emphasis added)

In thinking of affective tactility, following Gatens beyond conventional theories of the body will be useful. But her highlighting of Spinoza is also advantageous in another respect: we learn the language of *affect* from him, a term that does not deny the particularity of the body, but works to extend it. Linking affect with tactility will have resonance in therapeutic accounts, such as those of Reiki masters, and will show the relation between the sensory, experiencing body and the emotional character of the encounter in order to develop a 'felt' phenomenology, below.

Acknowledging that the body's limits and powers are indeterminable, we can go beyond this body as *topos*, as a site of in(tro)spection. To explore the mutual seepage of body and environment, to articulate those ambiguous energies that are of, around, within the therapeutic body, we can use the language of affect. Deleuze (1988:xvi, 123) and Deleuze and Gauttari's (1994:154ff) definition of 'affect' derives from Spinoza's *affectio* or *affectus*. It is not reducible to affection or personal emotion, but for Spinoza is rather the passage of a state as increase or decrease of potential power through the action of other bodies. As Massumi defines it, affect is 'a prepersonal intensity corresponding to the passage from one experiential state to another and implying an augmentation or diminution of that body's capacity to act' (in Deleuze and Guattari, 1988:xvi). Doesn't such prepersonal intensity trouble us, engaged as we are in establishing and celebrating the concrete particularities of bodies, and does this not contradict Gatens' reading of Spinoza? The point of thinking affect here is to examine the relations between bodies, not individual emotion. Affect goes beyond the 'attentional filter of representation' that 'seeks to capture experience as something inner, personal, subjective' (McCormack 2003:496). If emotion is the *personal* capture of feelings of intensity, then affect is unqualified intensity, an intensity that is actualised in the sensible materiality of the body, but which opens up this actualised intensity into something mutual between bodies, or between bodies and things, a passage between intra- and inter-corporeal intensities. It traverses the divide between the aesthetic, in the larger sense of the term as the sensory body, and the material, and encompasses those embodied feelings of tactility along with emotional feelings of touch – of being affected.

There is something disruptive about affect, however. Against the flow of highly disorganised and chaotic sensations in everyday life, as a perceiver I am able, seemingly negentropically, to organise them into ordered, recognisable perceptions. This is our usual mode of perceiving the world, our habituated perception. Affect often works antagonistically to this, argues Colebrook, since it 'disengages the ordered flow of experience into its singularities' (2002:24). It allows something to stand apart, obtrude, to reach out and touch us. It is to disrupt habituated perception through the force of altered, juxtaposed or disordered sensations. Rajchman likewise discusses this disruption of habituated perception, elevating it as the aim of art: 'through expressive materials, to extract sensations from habitual sensibilia – from habits of perception, memory, recognition, agreement – and cause us to see and feel in new or unforeseen ways' (2000:135). There are two interesting implications, leading to what in the therapeutic encounter we might term a *morphology of affective intensities*. That is, the exchanges and mutual transfers between therapist and client of emotional energies. Firstly, the ability to see and feel in new ways is applicable to therapeutic practices, where the transformative power of the therapeutic experience is both professed and sought after. And secondly, it eludes standardised forms of representation. In Margaret Boden's words it is 'enactive' rather than 'indicative' (2000:295), escaping from static, representational meaning by being based on interaction, and therefore from the singular, particular body to the interactions of multiple, mutable bodies. This concept of affect as the morphology and exchange of intensities is useful in thinking about therapeutic relations, and the imaginary of such relations through the words of Reiki practitioners and their clients, examined further in the next two sections.

The exchange of affective intensities between bodies is central to this analysis, since Deleuze and Gauttari conceive of affect as an autonomous being, having broken free from the agency of the human subject and the body, 'a bloc of sensations' as they say (1994:164). Affect is not just how an individual might feel, or what a particular body can do, but *what capacities it might have for relations with other bodies*. Illustrating this, and also underlining the ambiguity of affect, Deleuze argues: 'a body affects other bodies, or is affected by other bodies; it is this capacity for affecting and being affected that also defines a body in its individuality' (Deleuze, 1988:123). I want to argue that, at least in some cases, the therapist might be understood as a mediator of these capacities; a therapeutic stance can be reconceived as allowing these mutualities, potentialities and intensities to take place. This processing of a series of intensities, from one body to another, are affects. With the radical ambiguity of toucher and touched, the body conceived as a series of mutable intensities in this way allows interpersonal connection between therapist and recipient. By touching certain places in certain ways, one of the key points about Reiki and Shiatsu for example, it opens up the body's energetic pathways, allows the body to heal itself by opening up somatic energies into the larger field of cosmic forces. Thinking not of the private, delineated body as such, bodies become the capacity for affecting and being affected, of relations of speed and slowness, or intensities. This is effectively an ethology, states Deleuze:

> Concretely, if you define bodies and thoughts as capacities for affecting and being affected, many things change. You will define an animal, or a human being, not by its form, its organs, and its functions, and not as a subject either; you will define it by the affects of which it is capable. (1988:124)

However, before a detailed discussion of the actual experiences of touch and affect in Reiki, I wish to make a few contextual remarks about the links between touch, affect and space, with special reference to therapeutic spaces.

Touch and Therapeutic Space

The links between touch and affect could be analysed in a number of ways, such as through the skin surface. Massage, touching appropriately or inappropriately, and the complex feelings that arise such as the blush, which actually marks the skin itself through an affective state, is one way. Skin can be 'the "ambassador" of the psyche' in this case, argues Prosser (2001:58). But Reiki goes beyond the skin, trying to articulate touch in a wider sense than merely contact with skin surface. As such, this form of touch goes deeply into the affective realm, looking into the interiorisation of emotion and its subsequent exchange and release. As Elizabeth Harvey asserts, from a historical perspective, this form of touch lies 'at the interface between the psychic and the corporeal' (2003:15).

So far the analysis has hypothesised the exchange of intensities between bodies, and now I wish to consider the role of active movement in such an interchange. In therapeutic touch there is, and must be, both movement and proximity. Touching and being touched entails the movement of hands or body-parts, as well as the feeling, the cutaneous sensation. The sensory is bound to the sensorimotor, as that ambiguous touch of activity and passivity is equally spatial, requiring movement for sensation to take place. Merleau-Ponty confirms this ambiguity of the tactile, and we can extend this analysis into touch therapies. In his essay 'Eye and Mind' the body is not simply 'a chunk of space or a bundle of functions', but is 'an intertwining of vision and movement' (1964:162). Similarly, something of the tactile is intertwined with the kinaesthetic sense of movement in therapeutic practices: it necessarily involves both touch *and* movement. Touch as expression and gesture is both given and received. It is both active and kinaesthetic, through the act of reaching and touching, and passive and cutaneous, by being touched on the skin surface. Merleau-Ponty explores the meaning of tactility and visibility through the phenomenon of specularity that underlies them both. He suggests in *The Visible and the Invisible* that the 'fission' (*écart*) between touching and touched, seeing and being seen, actually reaffirms rather than undermines a strong sense of self. Merleau-Ponty's 'flesh' is therefore a mirror phenomenon:

> To touch oneself, to see oneself, is to obtain [...] a specular extract of oneself. I.e. fission of appearance and Being – a fission that already takes place in the touch (duality of the touching and the touched) and which, with the mirror [...] is only a more profound adhesion to Self. (Anonymous2000):255-6)

But this 'specular extract' affirms only the visual basis of a self/other boundary, is trapped in the conception of the body as a site of in(tro)spection. This is a trait of modernist models of mind, with metaphors of vision, reflection, and the self, even though Merleau-Ponty wished to move beyond this. Instead, dynamic touch is surely an intercorporeal phenomenon, affording a sense of (non-reflective, non-mirrorlike) mutuality. Touch is not enclosed self-identity, the model of vision that allows the 'repetitive affirmation of a closed ipseity' in the words of Vasseleu (1998:107), but is more proximal and opens into something literally quite *other*. It is not just that touch is a sense of nearness, whereas vision and audition are senses of distance. Touch is proximal in the way that modernist metaphors of visuality are not: 'Since empathy and sympathy are phenomena of proximity, they can only be understood as feeling-acts of a tactile rather than a visual subject', argues Wyschogrod (1981:32). Touch allows alterity by entering into a relation with another affective body, because it is empathic. Wyschogrod defines empathy as 'the feeling-act through which a self grasps the affective act of another through an affective act of its own' (1981:28), and her thinking of empathy through affect in this manner is an important way to think about what occurs in therapeutic practices. However, rather than a 'feeling-act', which suggests an individuated self performing an isolated act, I will talk of 'feeling-with', which invokes intersubjective, tactile empathy in a more unfolding, processual way.

Reminding us of the definitions of affect by Colebrook and Rajchman in the previous section, Tamsin Lorraine identifies the tactile especially as a form of 'disruptive excess' (1999:46), something that consistently eludes or escapes representation and habituated perception, and this must be the case for non-touching as it is with touching. In the absence of an adequate language for touching and feeling, I therefore proceed tentatively into an analysis of touch therapy, grasping towards a felt phenomenology of Reiki.

Reiki: Touch and Non-touch

There are many healing forms of touch, as Tiffany Field (2001:108) remarks, negotiated forms of touching, enabling intimacy, that proliferate even as our culture increases its restrictions on touching in public life. But there is a particular tactility that is implicated in Reiki massage, the guiding philosophy being that its therapeutic touch opens up the healing powers that the body already possesses. Touch is therefore crucial as it opens up a deeply affective energetic pathway. The link between this 'deep' touch of Reiki and emotional release (catharsis) will be explored here, drawing on my felt phenomenological experience of Reiki sessions as well as dialogues with Reiki practitioners. Reiki is a therapy of tactile encounter, yet includes non-contact aspects within its therapeutic practice. Feelings of well-being are amplified through this non-intrusive, therapeutic touch. It is founded on the allowance, the affirmation, and the confirmation of feelings that practitioners take into the sessions. It is held that the *Rei-ki* – 'universal life force' in Japanese – is passed between people through touch, and this makes the therapy an amalgam of tactile contact and non-contact, an inter-subjective energetics, what I have described above as the exchange of affective intensities. This is not simply another language of hypothetical abstraction, but given understandings by Reiki practitioners, is eminently suitable. Like acupressure and acupuncture, Reiki follows the principle of the internal life force as a form of internal 'aqueduct' system of flows in order for the body to function (e.g. Kaptchuk and Croucher, 1986:40). This life force or energy 'chi', or 'ki' in Japanese, is used for acupressure, acupuncture and tai-chi in Chinese medicine, and both Shiatsu and Reiki in Japanese medicine tie in the healer's energies with those of the recipient's through touch. These different traditions assume a system of energies which refer to 'the organizing factors that underlie what we call the life process', observes Montagu (1986:404), and he also acknowledges that non-contact in therapeutic touch can work to decrease anxiety. Reiki is perhaps more revealing than other touch therapies since non-touch is as significant as touch, the interaction with the body's energies arising both with and without actual skin contact, and the emotional release that often occurs at such a time. Attempting to understand this relation between contact and non-contact, massage and energy was central in my questioning of the practitioners. Touch becomes an irruptive, interpersonal experience that traverses skin and flesh, that cuts across Occidentally constructed divisions of mind and body, the outside and inside of skin.

On one level in Reiki there is an empathic touching, a 'feeling-with', the notion of touch as reassurance and comfort, of the interchange of affects. In nursing literature touch is another form of comforting and contact for healthcare professionals. It becomes assurance, based on empathy. The empathic content of touch comes from the patient, a sense of allowing the patient to appreciate that the nurse 'knows what it is like to feel as they do' (Edwards, 1998:815). On another level, the primacy of touching is a short circuit to feeling a deeper connection, feeling attuned to larger forces and energies. For example, Constance Classen reminds us that sensory orders are also already *cosmic orders*. They are not 'read' but 'lived' through the body (Anonymous1993a):127), referring beyond the actual site of the body in its concrete, sensed particularity. A sensory order is therefore not given as such, but culturally ordered. It refers beyond the body, but the hierarchy of the senses is variable between cultures and is reproduced through a culture's history. By connecting up with a cosmic order, the immediacy of touching opens out onto a much larger imaginary, one that accommodates the translations of forces and energies within an individual body into the vast networks of energy patterns that surround us. This is important to remember when considering touch therapies like Reiki. I want to illustrate how these themes of individual, empathic touch and the connection with a cosmic order work out through the stories of two Reiki therapists who practice in Bristol, Rachel and Louis. This is an opportunity to apply the language of affects and intensities to the ways that therapeutic touch is imagined and actually practised. These two practitioners represent two aspects of interest in this study, for while Louis talked more about the intensities and energy pathways, articulating something of the imaginary of cosmic forces, Rachel was more concerned with the emotive content of the Reiki experience, and talked of her emotional motivation to learn Reiki in the first place.

Louis is a healer who also teaches Reiki in workshops. He also uses Reiki massages as an introduction to more complexly codified massage such as Shiatsu. He articulates some of the themes mentioned, the notion of Reiki dealing within an affective economy that spills over and outside the body, not simply reducible to individual emotional states. He conveys the fact that in Reiki the 'aura', or what he calls the 'bio-energetic field', is something that can literally be felt, something actualised in a particular body but able to be exchanged between and beyond them. Exchange and transfer are facilitated through the touch and often non-touch of the therapist. He explained the interaction between bodies not as that of a more powerful body and a weaker body. Instead it appears more akin to a Spinozist body of energies and intensities, facilitated by exchanges between the healer's body and the recipient: 'So […] it's not a directive treatment, it's almost like you're turning a tap on for someone else to drink from', he explained, indicating a much larger set of surrounding forces and energies. The way that the body refers beyond itself, and his comment about the treatment as a tap, implies that the body is only a conduit, in his language a 'vessel': '[…] it's very much empowering the receiver, without having any sort of ego, which is one of the beautiful things about Reiki' he said (19/09/02).

The paths of energy, the exchange of affective intensities, are facilitated by the drawing of Reiki symbols on the body, working within an imaginary of a larger symbolic and cosmic order, as both Louis and Rachel described. There is a sense of connecting with something much more extensive, of proximity with larger forces. Speaking of her first experience of Reiki, Rachel describes how symbols were 'written' or inscribed onto her back by the masseur:

> I just lay down and this woman just started to put her hands in all different positions over my body, and I was feeling this tingling sort of feeling. And there was something magical going on here, I couldn't quite put my finger on it, I thought there's something going on here, something special. And she was drawing these little symbols into my back, these Reiki symbols, I didn't know what she was doing at the time but I could feel she was somehow almost *programming my body*, like this de de de de. And I felt that there was a part of me that understood, even if the rest of me was not really sure … [Rachel, 16/09/02]

Rachel's description of her first Reiki experience hints at a diagrammatics, a notion of gesture that articulates through motion and opens up paths for energetic and emotional transfer. She understands it as a coherent, interconnected system of symbols, of movements, gestures, and energies. The symbols serve as markers to channel the 'ki' energy, but this drawing out is not necessarily *on* the skin. It doesn't have to touch the body. A series of gestures, sweeps, motions, what Rachel showed me through sweeps of the hands ('*de de de de*', in the interview above), that participate in a mapping of energy through, around and between bodies: A morphology of affective intensities. After all, a diagram is 'the operative set of strokes and patches, lines and zones', says Deleuze (1993:194). It delineates and marks out areas of intensities and extensity. The use of symbols as a part of the therapy is integral for both practitioners. While the symbol has a distinct and 'correct' meaning when translated into English, it also simultaneously suggests a-signifying content, having 'magical' power; it summons, reaches back to, refers to something barely articulable through language, a set of pre-verbal sensations that is also thought of in Reiki as prior to being embodied. Louis for example says about the Reiki symbols that they are 'a vehicle for transfer'. In a portion of the interview concerned with the relation between touching and non-touching, this idea clearly came through:

> Louis: Well touch is […] very interesting, even though we're not touching each other sitting here now, we are sort of touching each other, you can touch someone with your eyes, or you can touch someone […] but might not actually feel a physical touching …

But with the response that this is still touching, he continued in a way that directly equates touching with the transfer and intention of energies:

Louis: Yes, if I lean towards you very quickly I'm touching you, and just in the same frame of mind, in Reiki we have the same experience, we have a bio-energetic field around us. Now various different analysts have said how far it goes. Some people say seven metres, some people try and quantify it, I don't believe there's a distance it goes, it's all down to your ... it goes as far as you want it to, as far as you send it. So the idea of prayer in ancient religions would be in bio-energetic terms a projecting of your energy field to *touch* somebody, to touch a situation – praying for world peace, praying for that – and in the same way distance healing or Reiki symbols or even if you're doing a hands-off feeling so you're giving Reiki but your hands are maybe six inches above the body, you're projecting your intention, the energy, you're visualising or using your 'intent' is the best word, to intend that that energy is going to be transferred or is going to be made available to that person if they want it. And the Reiki symbols ... are kind of like a vehicle, it's a vehicle for transfer let's say ... [19/09/02]

Within these exchanges and transfers of energy through the diagrams and symbols there is affective content; along with the physicality of (non-)touching there is often catharsis, release, sometimes crying. Rachel was in her early twenties, newly in therapeutic practice. She had her first encounter with Reiki at Glastonbury shortly after her father had died. Her identification of Reiki as a therapeutic practice with an explicit emotional content was made on numerous occasions. Explaining to me what would happen in the Reiki session to follow, for example, she said:

I'll put my hands in positions *on* your body, and also *off* the body, working on your aura, the sort of energy which is all around you, and I can feel that quite strongly [...] Sometimes you come out feeling a bit heavy, a bit disorientated, as if waking up from a deep sleep, but that's because it's aroused deep feelings and things, and people will go off and have a good cry or have a think, they'll just feel quite energised. *Something will shift*, then I'll just say drink plenty of water and take it easy, that sort of thing. [16/09/02]

The way that a tactile and, at times, non-tactile therapy can arouse deep feelings was surprising and unexpectedly powerful, as I found in an initial Reiki session. Halfway through my first massage, as described in the prelude to this chapter, there was an almost palpable sense of release. With my back to her, I was unable to see the movements of her hands, and due predominantly to the lack of actual tactile contact, trying to sense her movements, gestures, the marking out of symbols without directly seeing or feeling them was an uncanny experience. This sensation increased, became manifested as a great energy or intensity, and as physically-felt catharsis resulted in my spontaneously bursting into tears partway through the therapeutic session. For practitioners, this catharsis is a familiar occurrence. Early in our conversation Louis had identified the emotional content of the Reiki experience, for example, linking it explicitly with touch and the energies involved between bodies:

Most people experience a sense of relaxation, a sense of peace, a sense of nurturing. I think that's simply down to being touched [...] it's a subjective, non-invasive touch, but it's also a supportive touch, but as a result of the transfer of energy some people might experience a very beautiful, relaxing subsequent time, and *some people might have things stirred up, some kind of cathartic experience or emotional release,* depending on how they chose to respond or how they chose either at a conscious or unconscious level to use the energy that's been made available to them. [19/09/02, my emphasis]

From the practitioners' remarks I think there is an implicit notion of Reiki as being a form of touching that is not just immediate and physical, entwined with the particularity of the lived body, but which uses touch in a deeply affective way. In the imaginary of practitioner and recipient, personal intent and the manipulation of forces that are brought into proximity can be used to touch a person or even a situation. This I would argue is an example of 'deep' touch, a form of touching that is both intensive, that is affective and concerned with the sensation of the immediate subject, and extensive, occurring within a much larger imaginative framework of energies and events. Therefore, spaces of therapeutic touching allow what is distant and remote, these imaginations of larger frameworks, into the proximal immediacy of the encounter between tactile and non-tactile bodies; the mutuality of feeling-with.

Conclusion

> Touch is not a sense at all; it is in fact a metaphor for the impingement of the world as a whole upon subjectivity [...] to touch is to comport oneself not in opposition to the given but in proximity with it. (Wyschogrod, 1980:199)

The effectiveness of touch therapies such as Reiki perhaps lies in the preverbal interface between the psychic and the somatic, sensory experiences that lie prior to articulation. Their effectiveness is in pure affectivity; and this affectivity is stimulated and furthered by touch. This is a touch understood *ab initio* as a set of ambiguities: of the cutaneous contact and non-contact tactility that occurs in Reiki; of the 'deep' touch involved in touching and influencing people and events in the world, accessing pathways of energy through intention in the minds of the practitioners; of feeling and the felt, which is an active and a passive touching, while remaining affective or emotional; that, as such, these ways we have described of touching are all linked with affects, because to touch and be touched is to affect and be affected, in emotional, sensory or psychic life. What has been articulated theoretically, and actualised through the practice of Reiki, is the exchange of affective intensities, a form of touching that interfaces the psychic and the corporeal.

Van Dongen and Elema's plea for the consideration of the relations between the body, touch and the emotions in nursing (2001:153) is relevant to other areas of academic work, and I have explored these relations through the language of affect. Like affect, touch is ambiguous, and some of the ambiguity of these terms has pervaded this chapter. Thinking of touch, like Wyschogrod (1980), as actually a feeling of impingement upon and proximity to the world is helpful in therapeutic spaces. For to be affected, to be touched, is to bring aspects and forces of the world nearer to us. To go outside the body, to think tactility and touch outside the skin, we need to think of the relation between touch and affective experience. 'Sentience takes us outside ourselves', as Taussig (1993) remarks. This has added resonance when we think of the origin of 'sentience', the Latin word *sentire* meaning 'to feel'. We thereby return to the theme of an affective content to touching, an ambiguous tactility, a tactility that goes both ways, from the toucher to the touched, and back again. In the case of Reiki, feeling does indeed take us outside ourselves, in terms of the empathy involved in therapeutic touching but also the connection with much larger forces and energies. In the empathy of touching that is feeling-with, the surge of cathartic release as described is a way that touch can lead to the externalisation of an interiorised emotion, an exchange of affective intensities between bodies in an immediate therapeutic space. While it is an encounter between practitioner and client that refers to a much larger imaginary of forces and energies, it is possible simply that the empathic experience of feeling-with was enough to trigger the exteriorisation of emotion, the intense sense of cathartic release. Whichever, my felt phenomenology has shown the therapeutic space at that time as imbued with a deeply affective form of touching.

References

Anderson, Kay and Smith, Susan J. (2001) 'Editorial: Emotional Geographies'. *Transactions of the Institute of British Geographers.* 26:7-10.

Boden, Margaret A. (2000) 'Crafts, Perception and the Possibilities of the Body'. *British Journal of Aesthetics.* 40(3):289-301.

Bondi, Liz (2003) 'Empathy and Identification: Conceptual Resources for Feminist Fieldwork'. *ACME: An International E-Journal for Critical Geographies*, Vol. 2(1): 64-76. Online at http://www.acme-journal.org/vol2/Bondi.pdf [Last accessed: 14/06/04].

Colebrook, Claire (2002) *Gilles Deleuze.* (Routledge Critical Thinkers) London: Routledge.

Classen, Constance (1993) *Worlds of Sense: Exploring the senses in history and across cultures.* London: Routledge.

Deleuze, Gilles (1988) *Spinoza: Practical Philosophy.* Trans. R. Hurley. San Francisco: City Light Books.

Deleuze, Gilles (1993) *The Fold: Leibniz and The Baroque,* London: Athlone Press.

Deleuze, Gilles and Gauttari, Félix (1988) *A Thousand Plateaus: Capitalism and Schizophrenia.* Trans. Brian Massumi. London: Athlone.

Deleuze, Gilles and Gauttari, Félix (1994) *What Is Philosophy?* Trans. Burchell, G. and Tomlinson, H. London: Verso.

Edwards, Susan C. (1998) 'An anthropological interpretations of nurses' and patients' perceptions of the use of space and touch'. *Journal of Advanced Nursing.* 28(4):809-817.

Field, Tiffany (2001) *Touch,* Cambridge, Mass.: MIT Press.

Game, Anne and Metcalfe, Andrew (1996) *Passionate Sociology.* London: Sage

Gatens, Moira (1996) *Imaginary Bodies: Ethics, power and corporeality.* London: Routledge.

Harvey, Elizabeth D. (2003) *Sensible Flesh: On Touch in Early Modern Culture.* Philadelphia: University of Pennsylvania Press.

Kaptchuk, Ted and Croucher, Michael (1986) *The Healing Arts: A Journey Through the Faces of Medicine*. London: Guild Publishing.

Lorraine, Tamsin (1999) *Irigaray and Deleuze: Experiments in Visceral Philosophy*. London: Cornell University Press.

McCormack, Derek (2003) 'An event of geographical ethics in spaces of affect', in *Transactions of the Institute of British Geographers* 28 (4), Dec 2003, pp. 488-507.

Merleau-Ponty, Maurice (1964) *The Primacy of Perception and Other Essays on Phenomenological Psychology, the Philosophy of Art, History and Politics*. Evanston: Northwestern University Press.

Merleau-Ponty, Maurice (2000) *The Visible and the Invisible*. 7th Ed. Evanston: Northwestern University Press.

Montagu, Ashley (1986) *Touching: The Human Significance of the Skin*. 3rd Edition. London: Harper and Row.

Prosser, Jay (2001) 'Skin memories'. Ahmed, Sarah and Stacey, Jackie, (Eds.) *Thinking Through the Skin*. London: Routledge; pp. 52-68.

Taussig, Michael (1993) *Mimesis and Alterity: A Particular History of the Senses*. London: Routledge.

Van Dongen, Els and Elema, Riekje (2001). 'The art of touching: the culture of "body work" in nursing', in *Anthropology and Medicine* 8(2/3); pp. 149-162.

Vasseleu, Cathryn (1998) *Textures of Light: Vision and Touch in Irigaray, Levinas and Merleau-Ponty*. London: Routledge.

Wyschogrod, Edith (1980) 'Doing Before Hearing: On the Primacy of Touch', in Francois Laruelle (Ed.), *Textes pour Emmanuel Levinas* (Paris: Jean Laplace), pp. 179-203.

Wyschogrod, Edith (1981) 'Empathy and sympathy as tactile encounter'. *Journal of Medicine and Philosophy* 6, 25-43.

SECTION THREE
REPRESENTING EMOTION

Chapter 13

Ageing and the Emotions: Framing Old Age in Victorian Painting

Mike Hepworth

Introduction

Ageing into old age is an emotive issue but what exactly are the emotions of ageing? How are the emotions associated with ageing and old age given cultural frame and form? These questions concerning the age-related nature of the emotions are complex and may be explored sociologically from two interrelated perspectives: firstly by means of empirical enquiries into the personal or subjective qualities of the emotions experienced in everyday life by men and women as they grow older and, secondly, through an examination on the cultural level of evidence of social expectations concerning the emotions of ageing available in visual, literary and other sources. This chapter explores cultural evidence supplied by a small selection of Victorian paintings and the focus of interest is the images of emotions related to old age in these works of fine art. Sociological research into these two sources of evidence on the emotions associated with ageing suggests that there is tension between the subjective experience of emotions during later life and the images of the emotions which society considers to be appropriate in old age (see Milligan *et al.*, this volume).

This chapter is therefore in two parts: part one is an introductory discussion of the tension between the subjective and socio-cultural dimensions of the emotions of later life and its implications for a constructionist understanding of the issue. Part two takes the form of a cultural 'case study' of the construction and interpretation of the emotions of ageing in Victorian painting.

In the second part of the discussion examples are taken from the work of two popular British artists of the period: Luke Fildes' *Applicants for Admission to A Casual Ward* (1874) and *The Village Wedding* (1883), and Hubert von Herkomer's *The Last Muster* (1875) and *Eventide – A Scene in The Westminster Union* (1878). These examples have been chosen to display (i) the explicit 'framing' of the emotions of ageing within the painting and the organisation of space within the canvas, and (ii) to exemplify the complex interplay between the intentions of the artist, the artistic conventions of the time, and the response of the viewing public. The overall aim is to refer the wider question of the nature of the emotions of ageing to a specific examination of the construction of images which were deliberately designed during the Victorian period to stimulate a specific response in the popular imagination.

Emotions in Later Life

Emotional Geographies

One significant feature of emotional geographies refers to the social prescription, distribution, location and situating of human emotions. In other words, what society considers appropriate for individuals to feel in specific situations. The fundamental assumption is that emotions are interactive; the meaning and the quality of individual subjective emotional experiences are derived from processes of interaction between the individual and his/her surroundings and these surroundings include the available images of the day. Such a view is taken in Hochschild's pioneering sociology of the emotions, where the subjective experience of an emotion is understood sociologically as a bridge between the self and the external social world. In terms of individual experience, the expression of emotions may involve a tension between awareness of personal feeling and awareness of the expectations of others, which are in turn responsive to situation and place. In, for example, his recent discussion of theories of grief, Neil Small asserts the importance of language in this process. The 'interface', he says, 'between the social and internal worlds is negotiated via the medium of language, which constructs, and gives expression to, concepts of the self and the emotions' (2001: 41).

A good example of the complex structures of the emotional bridge between the self and culturally prescribed expectations is the social management of death in the Victorian family. In her detailed studies of the family papers of selected middle and upper class families, Pat Jalland (1996) shows that although the 'Victorian ritual of the deathbed and the funeral, as well as the language of consolation, were heavily dependent on Christian beliefs' (1996: 340), the experience of caring for the terminally ill, and for the dying themselves, could involve a heart-wrenching struggle between devotional prescriptions of the 'good death' and the agonising realities of terminal illness (compare Morris and Thomas, this volume). The result was, of course, an experience of emotional conflict in those whose illness or personal beliefs did not correspond with prescriptions of the 'good death' in the devotional literature. As Jalland puts it, '[i]n practice, the good Evangelical death required a rare combination of good luck, convenient illness, and pious character, and was achieved more often in Evangelical tracts than in family life' (1996: 38).

A 'convenient illness' was one that would leave the mind unimpaired so that the dying individual could consciously prepare for death and express hope of salvation through Christ. The ageing process could interfere with this when the mind was impaired. Jalland quotes Dr Samuel Beckett writing in the early 1850s that certain illnesses made it difficult if not impossible for 'candidates for immortal felicity' to engage in the appropriate verbal expressions:

In many cases, it is only in half-broken accents, or by the significant movement of the head or of the hand, that we can arrive at the assurance of the Divine presence and support; and the oppression of pain and disease may be so fierce, that even our inquiries seem to become almost an impertinence. Perhaps it may be truly affirmed that comprehensive and earnest views of life, death, and futurity, do not usually originate with such circumstances; apathy on the one hand, or nervous depression and irritability on the other, being far more usual. In fact, serious illness often becomes a touchstone of character; revealing weakness, instability, and impatience, where firmness and grandeur of demeanour might have been expected (Jalland 1996: 36).

Amidst these emotional conflicts one of the functions of religion was to provide consolation to believers and when a correspondence existed between expectation and experience, the 'good death' could undoubtedly be a positive emotional experience. 'After her father's death in 1886, Mary Booth concluded that, without belief in immortality, "life is so meaningless – and above all the closing years of it, so utterly stale, flat and unprofitable"' (Jalland 1996: 340).

But for those who found it difficult to believe in Christian salvation and the afterlife, the bridge between personal emotional experience and the expectations and beliefs of others was broken. As Jalland shows, the emotional experience of the death of a loved one took on quite different qualities for those who were avowed agnostics. She argues that because the Christian faith was so deeply entrenched in political, professional and scientific establishments, agnostics of the first generation (1850s onwards) felt particularly isolated when confronting death. Intense emotional difficulties were experienced by the agnostics Charles Darwin, Joseph Hooker (the botanist) and Thomas Huxley, who did not subscribe to conventional religious faith and could find no legitimate public source of consolation for the intense grief they experienced when their young children died. In this situation their grief was directed inwards and became almost incommunicable. Wilfrid Blunt, the agnostic who had lost his Catholic faith and dreaded death, described his situation as 'a solitude beyond the reach of God or man' (quoted in Jalland 1996: 339).

These examples display the problems individuals have with emotional experiences which they cannot articulate in the images of conventional discourse, and here a gender difference complicated the situation. The recommended way for men of Darwin's class to cope with grief was to retreat into work. When Hooker's 6 year old daughter Minnie died in 1863, Hooker wrote to his friend Darwin to tell him how he tried to drown his sorrow in his work because he felt that 'there is no other living soul with whom I can talk of the subject' (quoted in Jalland 1996: 351). Jalland comments: 'The archival evidence suggests that bereaved Christian fathers of the 1860s would not have found it quite so difficult to express their sorrow to friends and family' (1996: 351).

As Jalland (1996) observes, middle class women did not have the option of drowning their sorrows in work outside the home. Darwin, who worked almost entirely at home after his famous sea voyage in *The Beagle*, benefited from the constant emotional support of his wife and his sister-in-law, who encouraged him to share his grief and weep freely. Darwin's wife never lost her faith and the fate of her husband in the afterlife caused her considerable heart searching over the years.

Feeling Rules: Rule Reminders

This preliminary discussion of the complex relationship between personal emotional experience and culturally endorsed images of emotion brings us to Victorian paintings and their role in the construction of the emotions related to ageing. Paintings have been described as a 'visual language of emotion' (Hallam and Hockey 2001). In order to set the sociological scene, I'd like to relate this observation to Hochschild's concept of 'feeling rules' and in particular the notion of 'rule reminders'.

Hochschild's concept of 'feeling rules' (1985) is now regarded as a 'classic' contribution to the sociology of the emotions (see also Bondi, this volume). In *The Managed Heart*, Hochschild uses the concept to describe feeling as 'a script or a moral stance' toward action (1985: 56). The question she raises is one of how we become aware of the existence of these scripts or feeling rules. How do we identify a feeling rule and work through its relationship to our subjective feelings? – a problem she describes as 'the pinch' between '"what I do feel" and "what I should feel"' (1985: 57). 'Feeling rules are what guide emotion work by establishing the sense of entitlement or obligation that governs emotional exchanges' (1985: 56).

Hochschild asks how we recognise feeling rules and suggests that it is through the inspection of 'how we assess our own feelings, how other people assess our emotional display' and in terms of sanctions 'issuing from ourselves' and from others (1985: 57). For Hochschild the emotional experience is shaped by constant reference to 'rule reminders'. These reminders function with a range of social sanctions to shape the situational propriety of individual expressions of emotions. In this process, as indicated above with reference to death in the Victorian family, there is always the possibility of a gap between the 'ideal feeling' and the 'actual feeling' (1985: 61). In addition, there is always the possibility of 'misfitting' inappropriate feelings (1985: 63). And an important aspect of feeling rules is issues of timing and 'problems of placing' (1985: 76): 'Being in the right place ... involves being in the presence of an audience ready to receive your expressions' (1985: 67). The social characteristics of the audience around us significantly influences the emotions we publicly express. This process can be seen at work in Victorian art criticism and descriptions of the behaviour of viewers at art galleries from the 1850s onwards.

From the middle years of the nineteenth century art galleries became a vitally important source of collective visual experience which was increasingly mediated by art critics who rose to prominence during this period. At the Royal Academy in London, for example, immense excitement surrounded a new exhibition and vast crowds attended opening day. One significant cultural feature of the late Victorian period was 'the close relationship between ... artists and their viewing public' and the fact that 'art in the Victorian era reached a far wider range of social classes' extending from those rich enough to buy and commission paintings to those who viewed the works in cheap illustrated books, newspapers and periodicals (Gillett 1990: 192). When it involved a visit to an art gallery the visual experience was one of inspecting a painting and reacting to it in the company of others and under the scrutiny of many eyes. Extremely popular paintings like Herkomer's *The Last Muster* had to be railed off to protect them from public enthusiasm. Sociologically speaking, a visit to an open day was a social and emotional performance because the art gallery became a place where certain emotions could be publicly expressed. At second hand those who could not attend the actual exhibition could read art criticism which was itself often cast in highly emotive language. In addition, reproductions of popular paintings were widely circulated and available to an expanding public embracing all social classes.

The emergent band of art critics played an important role in the process of educating the emotions by reminding both artists and their public of the appropriate feeling rules associated with the framing of subject matter in paintings. One of the great arbiters was John Ruskin, who attached enormous importance to the emotional value of visual art. Critics were especially significant because the Victorian period was a highly complex era of challenging technical, cultural and social change and therefore ambiguity and an underlying sense of uncertainty. When times are uncertain, as Hochschild points out, 'the expert rises to prominence. Authorities on how a situation ought to be viewed are also authorities on how we should feel' (1985: 75). People look around for guidance on what to feel and how to act. There is also an hierarchical element in this process: 'in the matter of what to feel, the social bottom usually looks for guidance to the social top' (1985: 75).

As previously noted, Hochschild sees emotional experience as shaped by constant reference to 'rule reminders'. But another problem is that the rule reminders as such are not necessarily unambiguous and there is always the possibility of competing or conflicting interpretations. A very good example of the situated variability of feeling rules can be found in differences in the response to two paintings by Luke Fildes (1844-1927), namely his now famous exercise in 'social realism': *Applicants for Admission to a Casual Ward* (1874) and his later painting *The Village Wedding* (1883).

Emotions of Ageing in Victorian Painting

Rule Reminders in the Work of Luke Fildes

Casual Ward is based on Fildes' observation of a motley queue of desperately poor and homeless men, women and children waiting outside a police station for tickets of admission to the casual ward for a night's shelter. Fildes did not use professional models for this painting but indigent people he had observed during his walks through the streets of London. As such the painting is often classed as 'social realism': a representation of a small section of the street life seething behind the façade of Victorian prosperity. In this spirit, when *Casual Ward* was first exhibited

at the Royal Academy it was accompanied by a characteristically emotive passage
from a letter Charles Dickens had written about the poor, a few years before:
'Dumb, wet, silent horrors! Sphinxes set up against that dead wall, and none likely
to be at the pains of solving them until the *general overthrow*' (quoted in
Lambourne 1999: 137).

From the viewpoint of the crowd attending the Royal Academy Exhibition
this was an unusual painting. Fildes himself described the work as 'one to be
looked at and thought over' (Chapel 1993: 85), and recorded his anxiety over the
reception the painting would receive in a letter to his grandmother who had been
responsible for his upbringing and who had strong radical affiliations (Chapel 1993:
84). But he need not have worried; the painting was greeted with tremendous
acclaim and the police had to control the enthusiasm of the crowd by placing a
protective railing around the painting (Gillett 1990: 107). Art historian Paula
Gillett has noted that the appeal of this painting is to the emotion of 'compassion'
and also to 'the sense of social injustice ...' (1990: 118). And yet not everyone
agreed about the emotional content and the appropriate response it should invite
from members of the public. While the *Athenaeum* found the painting 'painful',
'morbid', not 'pleasant' yet 'far from repulsive' (quoted in Chapel 1993: 85-86),
the *Spectator* was not sure that 'mere misery like this is a fitting subject for pictorial
treatment'. The *Saturday Review* thought it was a mistake to choose such a subject
and a waste of talent, whilst the *Manchester Courier* found the figures 'repulsive in
the extreme', claiming that they 'simply appal and disgust without doing the slightest
good to humanity or making it more merciful ...' (Chapel 1993: 86).

Yet, at the same time most other reviewers were complimentary,
recommending a positive response, and discovering in the work a representation of
moral truth using the unvarnished techniques of realism. Fildes' painting found a
ready buyer and he never lacked a market for his 'social realism'.

Whilst *Casual Ward* is now treated as a sombre classic of English social
realism, Fildes' later painting *The Village Wedding* is considered to be a
sentimental genre painting and as such not in the same aesthetic league. This work
is a product of Fildes' post *Casual Ward* career as a member of the prosperous
Victorian art establishment. Whilst both these paintings include the figures of older
men and women there is a crucial difference between the cheery village labourers
in the *Wedding* and the down and out street people queuing for shelter in the
Casual Ward. The *Village Wedding* is a cheerful and colourful parade of the
stereotypical spectrum of an idealised village life. In the wedding procession down
the village street jolly-looking older rustics are an integral feature of an idealised
landscape. Unlike *Casual Ward* the colouring is unnaturally vivid and bright,
reminding viewers that this is the positive kind of response to the subject of a
wedding, in sharp contrast to the dark and forbidding colours in which a scene
from the everyday life of the Victorian poor is depicted. In an appreciation of the
art and life of Luke Fildes, published whilst he still had over twenty five years to
live, David Croal Thomson, editor of *The Art Journal*, passed the following
comment on *The Village Wedding*:

For years some of Mr Fildes' admirers had found his pathetic vein so strong as to be a little non-attractive, and he was determined to please everyone by painting the event which ought to be the happiest in the life of both man and woman. Here, indeed, was a change of subject, and the picture proved that the artist's admirers were not far wrong that he could paint happiness as well as misery (1895: 9).

In these excerpts from reviews of two of Fildes' works we can see that one of the functions of Victorian painting and art criticism was to remind viewers about the appropriate feelings a painting should arouse and therefore the emotional propriety of the subject. This is supported by the tone and language of art criticism of the period, which is nothing if not impassioned. Art critics became increasingly powerful during the period and their approval, as arbiters of public taste, could make or break an artist's career. The change of direction of Fildes' work as he became more successful is an object lesson in the influence of market values on a painter's choice of subjects and the style of representation, as Gillett (1990) has shown in her careful analysis of the world of the Victorian painter. Comparison of the public response to Fildes' two well-known paintings also shows that the communicative interaction between artist and public which is central to their rule-reminding role was by no means unambiguous, and space existed for rejection and reinterpretation as well as acceptance and confirmation. In other words, rule reminders only work if those to whom they are addressed ultimately wish to conform in their expression of emotions.

As indicated above the range of emotions critics informed viewers they should feel when looking at *Casual Ward* extended through compassion to anger and disgust. *Village Wedding* is associated with the emotions of amusement and delight allied with sentimentality, which was regarded by the Victorians as a positive and desirable emotion in both women and men.

Ageing and the Emotions in the Work of Hubert von Herkomer

As previously noted, in this chapter 'rule reminders' refer to the question of shared understandings between artist and their publics. I now wish to look more closely at two popular and still well-known paintings whose subject matter is the depiction of old age. These are Hubert von Herkomer's *The Last Muster: Sunday in the Royal Hospital, Chelsea* (1875), and *Eventide: A Scene in The Westminster Union* (1878).

The impulse behind both of Herkomer's paintings was deeply social realist. They were explicitly designed as companion pieces to record two aspects of old age, male and female, in two distinctive locations. The public reaction when the paintings were eventually displayed, however, indicated ambiguities in attitudes to representations of ageing which reflected deep-seated distinctions in the perception of the location, gender, and social status of older people.

Herkomer left behind a detailed record in correspondence, and also published during his lifetime extensive records of his aspirations and techniques. *The Last Muster*, a painting of Chelsea Pensioners gathered for Sunday worship in the Chapel of the Royal Hospital, was based upon a wood engraving he contributed to the realist illustrated paper the *Graphic*, 18 February 1871. Whilst composing and executing the work he described his progress in words which clearly indicate his interpretation of old age as 'a mass of old men sitting in their church during service ... There are about seventy heads to be seen, and all are literal portraits ... It is a grand sight to see these venerable old warriors under the influence of divine service' (Saxon Mills 1923: 87). The men were painted in his studio following the physiognomic principle that external appearance is a guide to internal character and moral worth in later life (Cowling 1989; Hepworth 2004). The idea was to record each subtle variation in individuality so that no man would blend indistinguishably into the congregation of old men. Each man was to be clearly identifiable, including his own beloved father, dressed as a Chelsea Pensioner. It is clear that Herkomer's blend of realism was closely in tune with one significant strand of Victorian aesthetic values. As Saxon Mills observes:

> There is no doubt that the circumstances of his life, his lack of brothers and sisters, the perpetual companionship of his parents, the spectacle of their long struggle with unrelenting adversity and his own acquaintance with the harsher aspects of life, had made him old beyond his years and given him a sympathy with toiling and suffering, and especially aged humanity, which is not common in young people (1923: 92-3).

Herkomer set out to remind viewers of the sympathy they should feel towards images of older people and *Muster* was an instant success with the public, catapulting Herkomer to a fame and fortune that lasted until the end of his creative life. 'It was said that members of the Royal Academy Council reviewing the works sent in for exhibition, "instinctively clapped hands" when they saw it ...' (Gillett 1990: 108). The positive reaction of *The Times* critic traced the source of its appeal to sentiments stimulated by two key features of the painting, which were the hand of death and patriotism (Gillett 1990: 108). The painting was also a success at the Paris Exposition of 1878 where it was commended for its explicitly English interpretation: although a foreigner, Herkomer had successfully penetrated the depths of national sentiment. In a letter to his aunt and uncle he shrewdly observed, '[t]here's hardly another subject that so appeals to English hearts – men who have fought for their country and have come to their last home preparing for their last journey home' (Saxon Mills 1923: 88). In the same letter he describes the subjects of his picture as 'all sitting, some with deep feelings of veneration, others more indifferent ... They have been loose (most of their lives), and now coming near their end a certain fear comes over them and they eagerly listen to the gospel' (quoted in Saxon Mills 1923: 87).

The positive reaction to *Muster* offers a fascinating insight into the ambiguous nature of Victorian attitudes to old age and the feeling rules surrounding later life. As we have seen, Herkomer was commended not because he painted old age as a neutral social realist but because he intentionally celebrated the heroic old age specifically of men who were growing older in one of the few respected institutions for the dependent poor. In the aftermath of the Crimean War, which had become a by-word for aristocratic incompetence, former rank and file soldiers came to be perceived by the public as representative of one of the forms of ideal ageing which the Victorians could reproduce and display publicly with equanimity (Hichberger 1989). Their distinctive scarlet uniform indicated patriotism expressed in combat and their worn and injured masculine bodies the signs, not of wayward self-indulgence or idleness, but of legitimate subjects for the artist, especially when framed in a spiritual location and 'living out a prosperous and contented old age' (Hichberger 1989: 55). As such the painting could legitimately associate the emotion of pathos with old age.

Figure 13.1 *The Last Muster* **by Hubert von Herkomer**

Source: Reproduced with permission of National Museums of Liverpool

The crucial difference between *The Last Muster* and *Eventide* is found in the fact that *Muster* is normally described as a painting about old age whilst *Eventide* has from the start been largely interpreted as a painting of poverty, and in particular poverty in a socially excluded location, namely the workhouse. In this respect it is useful to compare the response to *Eventide* with the reaction to Fildes' *Casual Ward*. As the subtitle of the painting indicates – *A Scene in Westminster Union* – the location is a real workhouse, a grim and predictably more realistic terminus than the Royal Hospital for poor people who lived long enough to reach old age. As such, the workhouse had been deliberately established to remind people of the virtues of hard work and to make them fearful of the consequences of an impoverished old age. But Herkomer had chosen this subject for subtly different reasons and saw this work primarily as a painting of old age. In a letter of 1876, Herkomer informed his uncle in America that he had begun working on a 'companion' picture to 'The Pensioners', as he called it (Saxon Mills 1923: 97), and *Eventide* was worked up from his double-paged wood-engraving *Old Age – A Study in Westminster Union* which appeared in the *Graphic* 7 April, 1877.

Figure 13.2 *Eventide – A Scene in the Westminster Union* by Hubert von Herkomer

Source: Reproduced with permission of National Museums of Liverpool

The immediate visual impact of *Eventide* is more muted than *Muster*. The scarlet of the Chelsea Pensioners' uniforms has a warmer effect than that of the black-garbed female workhouse inmates in their white caps. The perspective of *Eventide* is that of a stage sloping downwards towards the viewer so that the frame of the painting resembles the proscenium arch of a theatre. The whole drab scene is dimly lit by a large uncurtained window at the top left hand; a number of bowed figures whose features are indiscernible are seated in the distance and two others walk unsteadily in mutual support towards the foreground. Occupying a substantial space in the foreground, and illuminated theatrically from below, are six old women engaged in the conventional feminine pursuits of sewing and reading; their toothless sunken features do not, however, blur the individuality of their faces. The juxtaposition of youth and age as an emotive device for epitomising in static form the fluid passage of time is also evident: at the extreme right hand of the picture a young attractive woman assists the older women with their sewing and behind her on the wall hangs a reproduction of Luke Fildes' *Betty*, a painting of a young and beautiful milkmaid, famous in its day and first exhibited at the Royal Academy in 1875.

In composing this 'companion' painting Herkomer did not intend any explicit gender differential between the old age of men and women. Rather he set out to draw significant parallels between the experiences of the life of the poor as one of heroic struggle. Both the women and the men had fought the good fight and their features bore the marks of their endurance. Unfortunately the public did not react with the degree of acclaim that had greeted *Muster* and had established Herkomer as one of the most celebrated artists in Victorian Britain. It is the ambiguity of the public reaction which adds to our appreciation of the complex ways in which feeling rules influence emotional expression.

The reception of *Eventide* at the Royal Academy was far less favourable than Herkomer had hoped. The public did not respond positively to the parallels between the two paintings, showing that the artist was certainly out of tune with English feeling. The problem in England was the explicit representation in a sympathetic light of the workhouse as the last resting place for the dispossessed in order to invite from the public more positive emotions: to displace negative feelings towards older inmates of the workhouse and to generate positive feelings of veneration and respect. In this setting old age was not socially acceptable to many of the viewing public and there is the additional complication of gender. Feminist art historians have argued for a closer study of the ambivalence expressed towards women in Victorian painting (Cherry 1993) and Herkomer's painting of gendered old age is unusual because grandmothers in Victorian paintings are not usually located amongst the dispossessed of the workhouse. They are normally situated in family settings where they continue worthily to provide the family support of childminding, dispensing advice, cooking and sewing. The old soldiers in *Muster* were certainly impoverished and in receipt of charity but they were men who had earned their right in risking their bodies in patriotic conflict to the sympathy of those who were more fortunate. In other words, the bodies of old age were recognised and interpreted according to gender and the moral values attributed to specific locations. The setting of the Chapel (like that of the family hearth or meal) mediates the perception of old age as a moral process, the consequences of which are physiognomically displayed on the faces of the congregation.

Yet such is the ambiguity of the culture of ageing and the flexibility of feeling rules that there were contemporary viewers who thought the portrayal of ageing in *Eventide* sympathetic, and even that Herkomer had depicted a positive image of contentment in later life. Some critics like Baldry (1901) detected the universal quality Herkomer had wished to convey in his pictures of old age as 'observations of the ways of modern men' and 'dramatic episodes in human life which are independent of date or period' (1901: 53): the artist demonstrated a level of sympathy with older people which imparted a:

> deeper meaning to such portraits as 'My Father and my Children', and 'The Makers of my House', and to the many studies of aged types that he has painted. He has in all such works shown plainly that his choice of subjects was governed by something more than a love of picturesqueness; and he has distinguished them all with a seal of sentiment that is too fresh and unaffected to be other than persuasive to every one who is susceptible to the purer human influences (1901: 54).

In Baldry's interpretation a realistic treatment was justified not on the grounds of an appeal to a radical re-evaluation of poverty and social exclusion and a stimulus to political engagement with a world of structured injustice, but as an emotional reminder of the virtues of graceful ageing. And the preferred Victorian model of graceful ageing requires a resigned acceptance of the conventional demands of the preordained social station allotted to each individual. Herkomer's figures do not protest, as Baldry put it, in what is itself a significant catalogue of emotions, both positive and negative:

> with dramatic violence against a social system that grinds them down ... They suffer and have suffered; but they have accepted their lot with conscious courage, and have fought their fight with no questioning of its rights or wrongs. They neither rave against nature, nor whine about the injustices of the world in which they find themselves. Their dignity comes from the consciousness that they have done honestly whatever was entrusted to them, and that they have manfully fulfilled the duties, small or great, which were laid upon them as an unavoidable charge by a fate whose workings they could not hope to influence (1901: 54-55).

Conclusion

In this chapter I set out to draw on a small selection of popular Victorian paintings to explore one aspect of the geographies of emotions: namely the social prescription, location and situating of the expression of human emotions. I suggested that Hochschild's work on feeling rules and rule reminders provides valuable insight into this process and that the paintings under review can usefully be analysed as a historical source of information about feeling rules. I set out to show how the recorded history of art reveals a variety of responses to these paintings and therefore offers evidence of the flexibility of feeling rules and the spaces which exist for their variable interpretation. With regard to the emotions of ageing, this analysis substantiates the argument that, although society prescribes certain patterns of emotional response (for example, the negative emotions associated with the older women in the workhouse in *Eventide* or the bedraggled streetpeople in Fildes' *Casual Ward*) this does not mean that these rules are rigidly applied in the construction of social images or that they are slavishly followed by all individuals during the course of their everyday lives. As the efforts of the artists discussed in this chapter and the variability of the public response to their works clearly show, feeling rules are open to a number of interpretations at any given point in human history. But as an integral feature of emotional geographies they exist as essential cultural guides, or 'bridges' through the complex and difficult pathways to giving personal meaning to the experience of ageing and to establishing points of contact between subjective emotional experience and the attitudes of wider society to later life.

Acknowledgement

Some discussion of the work of Hubert von Herkomer in the chapter first appeared in J. Hughes and D. Inglis (eds), *The Sociology of Art*, published by Palgrave Macmillan (Basingstoke) in 2004. The author, editors and publishers gratefully acknowledge the permission of Palgrave Macmillan to include this material in this chapter.

References

Baldry, A.L. (1901) *Hubert von Herkomer: A Study and Biography.* London: George Bell and Sons.

Blaikie, A. and Hepworth, M. (1997) Representations of old age in painting and photography in A. Jamieson, S. Harper and C. Victor (eds) *Critical Approaches to Ageing and Later Life.* Buckingham: Open University Press.

Chapel, J. (1993) *Victorian Taste: The Complete Catalogue of Paintings at The Royal Holloway College.* London: Zwemmer.

Cherry, D. (1993) *Painting Women: Victorian Women Artists.* London and New York: Routledge.

Coveney, P. (1967) *The Image of Childhood.* Harmondsworth: Penguin.

Cowling, M. (1989) *The Artist as Anthropologist: The Representation of Type and Character in Victorian Art.* Cambridge: Cambridge University Press.

Croal Thomson, D. (1895) *Luke Fildes, R.A.: His Life and Work.* London: The Art Journal.

Edwards, L.M. (1999) *Herkomer: A Victorian Artist.* Aldershot: Ashgate.

Hallam, E. and Hockey, J. (2001) *Death, Memory and Material Culture.* Oxford: Berg.

Hepworth, M. (1998) Ageing and the emotions in G. Bendelow and S.J. Williams (eds) *Emotions in Social Life: Critical Themes and Contemporary Issues.* London and New York: Routledge.

Hepworth, M. (2004) Images of old age in J.F. Nussbaum and J. Coupland (eds) *Handbook of Communication and Ageing Research.* Mahwah, NJ and London: Lawrence Erlbaum.

Hichberger, J. (1989) Old soldiers in R. Samuel (ed.) *Patriotism: The Making and Unmaking of British National Identity.* London and New York: Routledge.

Hochschild, A.R. (1985) *The Managed Heart: Commercialisation of Human Feeling.* Berkeley: University of California Press.

Jalland, P. (1996) *Death in The Victorian Family.* Oxford: Oxford University Press.

Kern, S. (1996) *Eyes of Love: The Gaze in English and French Paintings and Novels 1840-1900.* London: Reaktion Books.

Lambourne, L. (1999) *Victorian Painting.* London: Phaidon.

Lowenthal, D. (1986) *The Past is a Foreign Country.* Cambridge: Cambridge University Press.

Marks, J.G. (1896) *The Life and Letters of Frederick Walker ARA*. London: Macmillan and Co Ltd.

Newall, C. (1987) *Victorian Watercolours*. Oxford: Phaidon.

Saxon Mills, J. (1923) *Life and Letters of Sir Hubert Herkomer*. London: Hutchinson & Co.

Sidlauskas, S. (1996) Psyche and sympathy: staging interiority in the early modern home, in C. Reed (ed.) *Not at Home: The Suppression of Domesticity in Modern Art and Architecture*. London: Thames and Hudson.

Small, N. (2001) Theories of grief: a critical review, in J. Hockey, J. Katz and N. Small (eds) *Grief, Mourning and Death Ritual*. Buckingham: Open University Press.

Yeldham, C. (1997) *Margaret Gillies RWS: Unitarian Painter of Mind and Emotion*. Lampeter: The Edward Mellon Press.

Chapter 14

Intimate Distances:
Considering Questions of 'Us'

Deborah Thien

Figure 14.1 *The Kiss* **(1916) by Constantin Brancusi**

Introduction

Brancusi's sculpture *The Kiss* (begun in 1907) illustrates a taken-for-granted contemporary story of intimacy: an intimacy we understand as developing usually, though not always, between two people as a consequence of love. The figures above are face-to-face, eye-to-eye, a visual and tactile representation of familiarity, closeness, understanding, relationship. Though the couple are not identical or exactly symmetrical, they make a satisfyingly perfect fit: two become as one. The embrace is clearly (hetero)sexual: visibly and differently gendered bodies press together in the act of a kiss, yet the intimacy on display suggests more than corporeal closeness. Due to the multiple meanings of 'touching' and 'feeling' (Sedgwick 2003),[1] touching and feeling bodies can also signal the nebulous world of emotions: 'I feel sad', 'I was touched by her kindness'. It is this type of touching and feeling, exceeding the corporeal, that has come to characterize contemporary intimacy.

In this chapter, I suggest that dominant contemporary understandings of intimacy have a particular spatial logic and that this logic recruits people to particular roles and rules, but that everyday lives also resist this logic.[2] I consider how the social proximity of Shetland's 'remote rural'[3] island communities in the North Sea highlights the permeability of boundaries (between, for example, senses of public and private), and how shifting interpersonal and community spaces of intimacy have varying consequences for people's sense of emotional well-being.[4] I propose that women's accounts of emotional well-being in Shetland unsettle some of the spatial assumptions of contemporary intimacy. Focusing on two such accounts, I suggest that the overlapping spaces of interpersonal and community encounters engender a more complex spatiality, and as such, both expose and challenge the underlying politics of contemporary intimacy. I conclude by considering how we might come to a different understanding of our intimate relations to, our being-with, others. This spatialised understanding might go some way towards answering the question 'shall we be, intimately and subjectively, able to live with the others, to live as others, without ostracism but also without levelling?' (Kristeva 1991: 2). This revisioning explores an alternative, elastic model of intimacy that incorporates the distances within intimacy in opposition to an emphasis on proximity. What I hope to articulate is an emotional geography of intimacy, demonstrating the ways in which intimacy is a spatial affair and how a spatial analysis of intimacy, something so inextricable from our sense of well-being, can enhance our understandings of emotional well-being.[5]

Spacing Contemporary Intimacy

The contemporary cultural narrative of an intimacy of disclosure has a distinct socio-spatial character, symbolized by the open arms of the embrace: 'I open myself to you' and its implicit companion 'I close myself to others'. The embrace marks the end point in a linear movement from distance (the space between you and me)

YOU $[$ --------------------------------------- $]$ ME

to proximity (being together as 'us')

$\coprod$

US

This intimacy assumes a distance covered, a space traversed to achieve a desired familiarity with another. As a vision/version of an achieved relationship (self to other), it is the antithesis of distance and as such the antidote to loneliness, unhappiness, estrangement and lack.

A willingness to disclose personal statements to another is a widely adopted measure of how 'healthy' one is, how 'emotionally honest', how 'in touch' one is with one's 'feelings'. 'Disclosing intimacy' requires the mutual and routine revelation of one's inner thoughts and feelings: 'It is an intimacy of the self rather than an intimacy of the body, although the completeness of intimacy of the self may be enhanced by bodily intimacy' (Jamieson 1998: 1). This relatively recent public story of intimacy as disclosure is deeply embedded in a contemporary perception of 'an individual's emotional well-being and of good relationships' (Jamieson 1998: 7). In western culture, we are inundated with demonstrations of such public disclosure in talk shows, reality TV, self-help manuals, and magazines[6] (Blackman 2004). This strategy for intimacy is characterized as 'being close' to another; as illustrated by *The Kiss*, it is an act of both touching and feeling that promises a union between two.

Within and through the rhetoric of disclosing intimacy, the space between two is filled in by knowing, until there is no empty space between (as *The Kiss* illustrates). As the knowing of each other and by implication the explication of the self increases, distance is transformed into closeness and two become as one. In this way, intimacy engenders an 'us' that is necessarily proximal and knowing. Operating as a sort of personal compass, as 'orientation and attachment' (Povinelli 2002: 231), intimacy allows for a bounding of the self and the (desired) other. In this way, intimacy is predicated on a high level of self-awareness:

> Intimacy, and personal sexual intimacy in particular, has come to be characterized by a form of pronominalized interiority. As numerous people have noted, the intimate interiority is characterized by a second-order critical reflexivity, by the I that emerges in the asking of the question, What do I feel towards you? In other words, the I who asks, What do I feel toward you? How do I desire you? contours the intimate interior. Along with being a form of orientation and attachment, intimacy is the dialectic of this self-elaboration. Who am I in relation to you? (Povinelli 2002: 231)

As proposed by Povinelli, the contemporary project of intimacy constructs a knowing self, an 'agentic voluntarist subject' (Blackman 2004: 231). The contemporary demand for a knowing self has converged with a heightened interest in a therapeutic 'working on' the self, a process that is taken up in a number of ways, including through the practices of self-disclosure, self-help, and counselling (see Rose 1990). Depression, popularly understood as a lack of well-being characterized by social isolation and the feelings and practice of loneliness, is the antithesis of this knowing self, producing instead a figure of *disorientation* and *detachment*. In a meditation on love, bell hooks writes: 'Although we live in close contact with neighbours, masses of people in our society feel alienated, cut off, alone. Isolation and loneliness are central causes of depression and despair' (hooks 2000: 105). Union with others, becoming part of 'us', is perceived as a step towards well-being.[7]

Buried within the promise of a union between self-aware subjects is the presumption of equality (Giddens 1992). In the liberal humanist framework which has infused late twentieth century western culture, individuals are understood to have agency and creativity in their self-production and self-narration. Factors such as gender, race, age, bodies and economic (in)security are seemingly subsumed in a process that emphasizes the rational production and narration of 'I' (Jamieson 1998). As Williams (2001: 91) remarks: 'Today for the first time, we are told, men and women face each other as "equals", intimacy holding the potential for true "democracy" not simply in the privatized domestic sphere, but also with the broader body politic'. However, research indicates that such equality remains an ideal, not a reality. Jamieson (1998: 164) notes: 'The thesis that couples are increasingly centred on disclosing intimacy suggests that it is theoretically possible for a couple to bracket off the material, economic, and social aspects of their relationship; whether this is theoretically possible or not, there is no clear evidence that it is happening in practice'. Furthermore, this ideal is in itself politically suspect. Young (1990: 309) critiques this ideal of the 'unification of particular persons through the sharing of subjectivities' as it is articulated within the project of community:

> People will cease to be opaque, other, not understood, and instead become fused, mutually sympathetic, understanding one another as they understand themselves. Such an ideal of shared subjectivity, or the transparency of subjects to one another, denies difference in the sense of the basic asymmetry of subjects.

Psychoanalytically-inspired treatments of subjection challenge the ideas that subjects can 'cease to be opaque' or that they can 'understand one another as they understand themselves'. Judith Butler's (1997) reading of subjection highlights how subjects are not so fixed, nor so wholly, freely or rationally self-producing. Butler extends Foucault's theory of power to explicitly address conscious and unconscious processes of subjection. Importantly, she explores how subjectivities are self-constituted, emphasising how the psychic and the social circulate through the production and reiteration of subjects. Theoretically and methodologically, this insight acknowledges that subjects are always partial in their presentation, and as Probyn (2003: 293) also argues, subjectivities are 'always' conducted *in situ*. Furthermore, the psychic and social reiteration of subjectivities allows for the possibility that we may be 'strangers to ourselves': 'a symptom that precisely turns "we" into a problem, perhaps makes it impossible' (Kristeva 1991: 1).[8] Kristeva (1991: 2) poses the question thus: 'shall we be, intimately and subjectively, able to live with the others, to live as others, without ostracism but also without leveling?'. Kristeva makes the point that we must recognize the stranger within ('to live as others') in order to accept the stranger without ('to live with the others').

In order to illuminate and illustrate the issues at stake, I turn now to two empirical accounts of living 'intimately and subjectively' within the relations and proximities of Shetland. In the discussion that follows, I am interested in how women on whose stories I draw reflect on and engage with the expectations, pressures, and pleasures of intimacy.

Living with(in) Intimacy

Given the strongly held assumptions of contact and familiarity within the social and spatial story of intimacy, it is unsurprising that the rural locale should be viewed as the place, par excellence, of warmth and close association. At the base of this assumption is a simple calculus: the social proximity of life in a geographically distant (from others) place is equal to a correlated emotional closeness amongst the inhabitants ('us'). Hester Parr (2002) reflects on rural Highland communities: 'people are often physically distant from neighbours (particularly in crofting communities) but more socially connected than in urban localities'. Yet, as Jo's story illustrates, the reality is both more and less than this equation (see also Parr *et al.*, this volume).

Jo is a slender, nervy woman, an incomer halfway through her second decade in Shetland. A business woman with young children, Jo's schedule is hectic and perhaps not surprisingly, her idea of well-being emphasises peace and stability, 'a new way of being which is just to be quiet'. She elaborates:

> [Emotional well-being], it's not anymore about just being really happy. I don't expect that anymore. I think that's an unrealistic aspiration for us all to have in life just that we're [the modern generation] all going to be blissfully happy ... I think looking back a hundred years ago people just had different expectations of life there was much more disease and death and illness and hard work, people just didn't expect so much and I think, this last century we've all come to expect so much and advertising's part of that. Everybody expects this glorious pastel white life, happy families and everything happy and of course that's very disappointing 'cause life isn't like that.

Jo's philosophy for achieving that peace is to reject the blissful happiness of advertisements, a 'glorious pastel white life' peopled with 'happy families'. She recognises contemporary intimacy as a public story 'offering stereotypes and ideals rather than the details and contradictory complexity of real lives' (Jamieson 1998: 159). Jo's critique is suggestive with its intimations of cleanliness, transparency, and an assumption of the raced[9] nature of happiness. In Jo's view, happy families are the imaginary product of commercial and social promotion of normative values. Trying to match life to this image, Jo concludes, leads to disappointment, 'cause life isn't like that'.

Despite this stoic statement, Jo struggles to keep this perspective in place. The recent dismantling of her own family, caused by the 'life-changing moment' of her husband leaving her, has led to Jo 'totally reassessing' herself. The breakdown of a central proposition of contemporary intimacy ('us') is painful. Jo finds she is placed outside of familiar relations. As a single person, Jo has lost the unremarked/unmarked subject position she was afforded as part of a marriage. As a single woman, and a single mother, Jo has a heightened sense of her visibility and feels keenly the loss of her previous status as part of 'us'.

As Jo sketches her life as a single parent, it becomes clear that a significant consideration in her life is the proximity of her neighbours. Jo lives in a small village and, as in many Shetland communities, her home is highly visible to those who live around her. It is this very kind of proximity in rural and remote communities that is usually perceived as a benefit. Close neighbours are assumed to equate to communal, neighbourly relations and for a newly single parent it might be assumed that proximate neighbours might ensure help and support with childcare. However, this is not the angle that Jo presents. For Jo, this proximity translates into a series of socio-spatial constraints on what could be termed her 'private life'. Contrary to any expectations of privacy, Jo's presence or absence at home is easily noted and potentially remarked upon, and so are her movements in or out of the house. During our interview, she tells an illustrative story:

> Some friends went out on Friday night ... and I couldn't join them. The next day, one of them popped in and told me how the evening went and the general discussion ... They were talking about *me* which is normal, and that's just normal, discussing was I seeing anybody, was I shagging somebody ... I'd been seen getting out of a man's car [outside my house] so then this discussion had ensued about was I having a relationship with this person who had given me a lift [laughs and throws up her hands] and that's just what it's like. You don't have to do anything here to be talked about ... I wouldn't be surprised if by next week it's fact that I'm having a relationship [laughs].

The monitoring of this encounter which took place outside her home is quickly translated into gossip which just as quickly circulates back to Jo. In a small place there is a certain acceptance of this process. Speaking of another island community (Antigua), the writer Jamaica Kincaid (1988: 52) remarks: 'In a small place ... [t]he small event is isolated, blown up, turned over and over, and then absorbed into the everyday, so that at any moment it can and will roll off the inhabitants of the small place's tongues'. Indeed, Jo emphasizes the everyday nature of this attention to the small things and she does not suppose there is malicious intent. But, she also conveys a wry chagrin. From her perspective, 'nothing' has occurred. But, she does not have to communicate an explicit romantic intent in order to be subjected to speculation. Proximity is readily interpreted as intimacy, thus the man in the shared space of a car that is also outside her home (the heart of intimacy) is assumed to be an intimate. Friends who have no other information can still 'read' this scenario as a potentially intimate encounter because of these socio-spatial arrangements. In fact, the man in question had simply given her lift home when her own car had broken down: a situation that owed more to mechanics than to ardour.

Jo sums up the (mis)interpretation of this scenario:

> It's the fact this was outside my house, which kind of presumably signals to the person who saw me that we'd been somewhere together which it wasn't like, it was just that I had been having all this trouble with my car ... so I got a lift [laughs].

The man in the car has crossed into her personal space; a personal space which is also a highly public and monitored space. While Jo can laugh at this state of (a lack of) affairs, there are consequences. She is highly aware that any desired, intentional interactions will undergo the same scrutiny and surveillance and she is afraid of the pressure this places on her. Jo feels she has no space to make mistakes:

> It's not like south where you could just anonymously go out with someone once or twice or once, never see him again that's the end of it. Here if you go out with somebody once you're taking on [the fact that] everybody's thinking you're together ... Now for me I don't want to go out with somebody unless I really really *really* like them and I would be really sure that there was something in it.

The pressure of practising intimacy in the public eye is so great that Jo is prepared to forgo the chance unless she is certain there will be 'something in it'. She does not want to risk publicly failing unless there is something worth risking. Unspoken, but understood, is her fear of failing again. The loss of her relationship with her husband has brought Jo face-to-face with the fragility of contemporary intimacy.

Aside from her anxieties of finding herself in another relationship that could break down, Jo has to consider her reputation as a respectable woman and as a mother:

> It's no longer just a case of I can do anything I want, I can just go shag a different man every week ... it's not like that, I have three children to consider, I have my neighbours watching me, I have an ex and we're not divorced so, I'm still married anyway so, chhh [shrugs].

As a newly single woman living in a small community, Jo must face the potential repercussions of any gossip for her or her children: 'I felt very concerned about my reputation, for myself and also for my children. I didn't want people laughing at me and talking about me ... people gossip a lot around here so my reputation became very important to me'. Her only option for dating, she feels, is to escape the proximity of her community:

> It would be a very difficult thing to actually go on a date with anybody secretly, unless I went south and you'd have to go even further than Edinburgh because if you try to go on a date in either Aberdeen or Edinburgh you'd bump into somebody from Shetland, so you'd be seen. If you wanted to [go] on a [date], I'd have to go to London.

Achieving the necessary distance from her highly proximal social world requires extreme action. A trip to London is no small expense of time or money. Her lack of 'fit' into her previous coupled state, her visibility as 'uncoupled', and the resulting consequences for her and her children leaves her feeling fragile and paradoxically, given her surroundings, isolated.

Facing up to Intimacy: Strange(r) Proximity

Such is the strength of the belief in the peace and bonhomie of rural living that well-being and other issues of mental health in rural communities have had little attention, other than to make assumptions of the 'good life', especially in a British context (but see Milligan 1999; Parr 1999). In fact, where mental ill-health is concerned, respondents in one Scottish study reported few socially supportive possibilities in their rural environs (Milligan 1999). Drop-in centres, transient environments by design, provided their best option. Of particular interest here, is how the 'perception of stigma and heightened visibility in rural locales contributes to create geographies of [mental ill-health]' (Milligan 1999: 230; see also Parr *et al.*, this volume).

Judy is about to turn forty but looks closer to thirty, in marked contrast to the weather-beaten, wind-blown stereotypical image of the northern woman. Judy says her youthful appearance is all in the skin – her mother and her grandmother had it as well. Judy describes herself as 'full Shetland'. Though she now lives in Lerwick (Shetland's largest community), she doesn't consider herself 'Lerwick', as her mother and her father's people both came from the South End (of the largest island), making them South-Enders, not 'sooth-moothers'. This dialect word 'sooth-moother' refers to those who come via the south mouth of Lerwick Harbour, the historical entry point of strangers into Shetland. The time she spends marking this difference to me signals how critical distinctions of place are, and, where 'sooth-moother' translates to incomer, also signals how places are constituted by one's positioning in a network of social relations.[10] Judy, in other words, is no stranger to this community.

In our conversation about emotional well-being, Judy tells me she needs to get away from Shetland regularly to keep herself healthy: 'I actually like going somewhere and I know that nobody's gonna come up and ask me something like or know who I am'. She says 'it can get too much' with everyone knowing her and knowing everyone. In the following interview excerpt, Judy describes her unease with being inescapably familiar:

> Judy: I was meeting a couple a friends, we meet at [a local café] at half past five and have a pot of tea. They didn't come and I was looking at me watch and thinking I'm sitting there drinking tea at [the café] on my own, how strange. And I started to feel, there's people looking at me [laughs] ... they turned up about twenty minutes late and I felt really uncomfortable. If it had been anywhere else Aberdeen, Edinburgh, it wouldn't have mattered, I'd be sitting there quite happily, peace and quiet [laughs]. [But here], somebody could pop their head in and say 'Judy what are you sitting there yourself for?' Somebody might say that to you.

Echoing Jo's experiences, the proximate public spaces of Judy's community result in an intense visibility and this does not provide Judy with a sense of well-being. Rather, she is 'really uncomfortable' and feels 'strange'. Paradoxically, it is precisely because she is *not* strange, because she is in fact recognizably *familiar* that she feels ill at ease. In part, this is because this meeting in this café presages different meetings in different places (Ahmed 2000).[11] Judy is aware that because she is familiar but behaving strangely (by sitting alone in a café), she may have to account for her actions (in a way she would not have to in other strange(r) places).

This awareness of her visibility and the potential for social stigma therein, comes up again when Judy, smiling, animated and confident, wistfully confesses to an unfulfilled desire for a bike:

Judy: You see I would actually love to get a really good bike ... but I just feel I'd look that stupid on it here. That's a thing I feel that what would, 'what's Judy Johnson doing on that bike' y'know. I didna have the confidence to do it. Whereas if I was somewhere I didn't really know people I'd get on the bike and off I'd go.
Deborah: Yeah, you don't think you'd get on the bike and people would say 'oh what a good idea Judy has'. [laughs]
Judy: I don't know, I don't know, I'm thinking aboot it, but [laughs] that's how I feel
Deborah: Uhhuh, that it would stand out.
Judy: Y'know, people do spaek[12] here, y'know they talk quite a bit ... if you do something slightly different here then folks spaek about you.

Again, Judy feels unable to behave strangely. She cannot take the part of the strange(r) because she is so emphatically not strange. She conforms to what is expected of her and thus claims her place within the intimate confines of the Shetland community. But, she also describes a mingled sense of regret, resentment and resignation at this subjection ('folks spaek about you').

In Goffman's (1969) analysis, social actors define and bound particular spaces so as to manage discrepancies or disruptions to a coherent performance. Judy's arrangement to meet her friends at a designated time in a particular cafe is a way to bound the space of their meeting, however, this is disrupted by their late arrival. Judy is left feeling anxiously aware that she is out of place. In Goffman's (1969: 11) terms, she is experiencing the potential anomaly that occurs when performances are contradictory, wrongly defined or undefined and the result is a breakdown of face-to-face interaction: 'At such moments the individual whose presentation has been discredited may feel ashamed while the others present may feel hostile, and all the participants may come to feel ill at ease, nonplussed, out of countenance, embarrassed'. Goffman (1969: 58) proposes that a means to limiting and regulating what is available for perception by others is to manage the distance between performers and audience through the controlled use of information in particular spaces. People seek to avoid such disruption, as Judy does by living without a bike, in order to eliminate uncomfortable emotions such as embarrassment or shame, or the discrediting of an individual's self-conceptions (Goffman 1969: 219). The potential for intentional and unintentional disruptions (or misrepresentations) in this process of expressing the social self are processes of both recognition and misrecognition: 'a kind of information game: a potentially infinite cycle of concealment, discovery, false revelation, and rediscovery' (Goffman 1969: 6). Goffman's paradigm thus offers some insight into why Judy feels so uncomfortable in these situations. In the case of the café, she is on stage but without the correct props (the friends who are late for tea). Her performance is disrupted and she has no means to control the feared perceptions of her audience who will misrecognize her as alone in a café.

Facing up to others in such close proximity does not lead to a comfortable intimacy, even though the amount of known information may be increased. Judy offers one strategy for managing the demands of such daily encounters by maintaining an intimate relationship with a friend far outside her community. This allows her to stay in place, without sacrificing her need for confidences, or testing the limits of a tightly bounded 'tight-knit' community. Though Judy herself doesn't make this connection explicitly, her relationship with a life-long pen pal who lives overseas struck me as a very effective way both of 'getting away' from Shetland and getting around the issue of trust and of the associated visibility of proximal spaces. Although Judy later notes she has a 'best friend' in Shetland, and also describes herself as telling most things to her partner, she describes this long-distance friendship as something just for her, something private. Describing the relationship, Judy comments:

> I know that she tells me things and she trusts me ... she has done quite a lot, 'specially with difficulties in her life and I've done the same back to her so, I do feel that I do know her.

The two women have shared 32 years of letters and, in a time when emails offer instant gratification, Judy says she has kept every one. As aware as she is of the limitations of living where everyone knows you, Judy strategically expands her capacity for intimacy by stretching her relations beyond the limits of the community's purview. This relationship is an effective vehicle for practising an intimacy without the obstacles that the small community engenders.

The Contours of Intimacy

The accounts offered by Jo and Judy illustrate some of the contradictions of living within and facing up to contemporary intimacy. For Jo, the conversation we shared about emotional well-being was told through her recounting of the breakdown of her marriage and her subsequent efforts to rediscover stability, if not happiness. As a single mother, she is subjected to a set of moral codes about respectability and femininity within the social spaces of her community. Her search for intimacy in the form of a love relationship is hampered by the proximity of her social community and this is not restricted to the literal spaces of Shetland. Jo's fears and anxieties about making a new relationship are compounded by the limited personal space she can operate within. While Jo to some extent accepts the boundaries of her community, she also longs for some distance from it. Instead of intimacy being a function of proximity, the proximal spaces of her community are limiting her access to an emotional freedom she is currently without. Though Jo has concluded that 'life isn't like that', she still hopes for a partner to share her life with. It is this underlying hope that makes the complications of her socio-spatial situation difficult, exposing the contradictory feelings, distances and dilemmas of intimate encounters.

Judy's choice of a geographically distant friend does not match the cosy familiarity promised by the rural idyll, nor does it match the expectations of intimacy as proximal. Instead, Judy's heightened consciousness of her visibility within a socially proximate community negatively affects her sense of well-being. Her well-being is enhanced by maintaining a relationship that extends outwith her community and within the self. Furthermore, Judy's feelings of unease are not simply about what is known or not known of her, nor simply about whether she can manage actually occurring situations; they are also about how imagined potential encounters ('somebody might say that to you') subject her to the stranger within.

Stretching Intimacy through Difference

I wish to question the popular contemporary version of intimacy as the democratic and mutual disclosures of knowing selves because this is the version that many of us compare our lives to, and invest emotional energy in, even if we do not or cannot practice it (Jamieson 1998; Blackman 2004). This intimacy implicitly assumes the coming together of selves that are constant, stable, self-enclosed; selves that are gendered only in that the feminine forms the other half of the masculine same (two become one); selves that are both knowing and known – at the opposite end of the line to anonymity's unknown stranger. *The Kiss* is only able to project the solidity of two as one within this masculinist rhetoric of constancy and equality. As has been written about the desire for (certain kinds of) community, it entails 'the extrusion of alterity, in order to bask in the warm glow of self-confirming homogeneity' (Morley 2001: 441). Luce Irigaray (2000) has argued love must resist such fusion, and I follow this by suggesting that intimacy must necessarily incorporate alterity. If intimacy is understood to involve unstable and/or strange selves 'as others', it can be read differently such that distance does not separate in the same way, and neither does proximity (simply) bind. These different understandings of intimacy matter when it comes to considering how we operate in place, how we are differently gendered, how we place our relationships, how we think through them, enact them and desire them to be.

The empirical examples suggest that intimacy can be extended beyond the usual co-ordinates of 'inner' or 'close'. This move reflects an 'abandoning [of] the fiction of natural space' (Callon and Law 2004: 3) with its assumptions of correct positions and fixed opposites, and a taking up of a more fluid process. Intimacy as distance, intimacy as difference, offers a flexible intimacy, reflecting the ambivalent and elastic spatialities of women's emotional landscapes. The narratives elaborated by Judy and Jo about their emotional well-being and how they negotiate and perform intimacy in the particular setting of the Shetland Isles offer some alternative readings to dominant presumptions of contemporary intimacy. It is my contention that the story of intimacy as equalling a democratic proximity can be unsettled by their particular and sometimes contradictory experiences of being in the world.

An examination of the spatialities of contemporary intimacy reveals the limits of its contribution to well-being and unsettles assumptions about intimacy and its relation to proximity and distance. Spatialities of intimacy are revealed as not simply about the closeness of 'us' in polarized opposition to a strange(r) distance. The intimacies of our daily lives are not the product of a linear movement from distance to proximity, as a contemporary reading of intimacy would have it. However, intimacy is a spatial affair. As the above examples demonstrate, psychic, social and spatial relations ensure that distance and proximity can co-exist and may be configured in complex ways as we engage in our lives. Intimacy is also not simply the prerequisite for well-being as the popular story of intimacy suggests. The practices of intimacy, itself a paradoxical affair, engage unequal and gendered subjects and these engagements may produce ambivalent consequences, as well as positive affects. Considering the practices of intimacy with an awareness of the ways in which 'we' can be both knowing and unknowing, the ways in which 'we' are gendered and placed works to de-fuse the 'us' of contemporary intimacy. By examining subjective experiences of intimacy, intimacy is necessarily stretched, pushing us to live 'as others' (Kristeva 1991). This insistence on making use of difference incorporates the multiple and elastic experiences of intimacy which contour our intimate distances, and so the spatialities of intimacy are redrawn.

Notes

1 In Eve Kosofsky Sedgwick's book, *Touching Feeling*, she notes her 'intuition that a particular intimacy seems to subsist between textures and emotions. But the same double meaning, tactile plus emotional, is already there in the single world "touching"; equally it's internal to the word "feeling". I am also encouraged in this association by the dubious epithet "touchy-feely", with its implication that even to talk about affect virtually amounts to cutaneous contact' (Sedgwick 2003: 17).

2 This research is part of my recently completed doctoral research into women's emotional well-being in Shetland, for which I am most grateful for the support of the Commonwealth Scholarship Commission (Thien 2005).

3 The Scottish Executive has developed a 6-Fold Urban Rural Classification. The designation 'Remote rural' is applied to any community that is outside a thirty minute drivetime from the centre of a town with a population of 10,000 or more.

4 While investigations into the respective realms of personal relations and community politics are often pursed in separate studies, I approach these realms as overlapping areas. From my (feminist poststructuralist) perspective these realms bear a 'family resemblance' (Wittgenstein 1953; see Davidson and Smith 1999 for a further discussion of Wittgenstein's use of this term): namely, within the British western context of this research, both personal *and* community relations are infused with the same masculinist ontology. In this ontology, gender is based on a masculine same and a feminine other, in contrast to, for example, feminist revisions of gender as relational (e.g. Irigaray 2000).

5 The complexities of social and spatial distance are an abiding preoccupation within social and cultural geography (e.g. Massey 1997), an interest that can be traced in part to the early twentieth century urban sociology of the Chicago School (e.g. Park, 1926). Developing the Chicago School's symbolic interactionism, Erving Goffman (1969) famously sketched detailed observation of face-to-face encounters in the Shetland Isles in an effort to understand how people perform socially in the various regions of everyday life.

6 The self-help culture of magazines, Lisa Blackman suggests, is 'peculiarly feminine' and leads to the 'cultural production of female psychopathology' (Blackman 2004: 220).

7 In opposition to the nineteenth and early twentieth century practice of housing the mentally ill in asylums, the current belief that community care is a more inclusionary means for assisting those with mental health needs demonstrates that achieving a place of union with others, becoming part of 'us', is now perceived as a step towards well-being. Though, in fact, studies suggest that people with mental ill-health remain socially excluded even with the proximity of their communities (Milligan 1999).

8 An aspect of understanding psychic reality is recognizing how we are already 'other' to our selves, as Kristeva (1991) argues. In *Strangers to Ourselves*, Kristeva advances the claim that the desire to close the gates on the foreigner arises from the anxiety of facing one's own difference.

9 For geographical explorations of the social construction of 'whiteness' see Kobayashi and Peake (1994) and Jackson (1998).

10 Anthropologist, Anthony Cohen (1987) reports a similar significance given to genealogy as a method of mapping social knowledge on the Shetland isle of Whalsay.

11 As Sarah Ahmed describes (following Goffman): 'This [face-to-face] encounter is mediated; it presupposes other faces, other encounters of facing, other bodies, other spaces, and other times' (Ahmed 2000: 7).

12 This Shetland dialect word translates to the English 'speak'.

References

Ahmed, S. (2000). *Strange encounters: embodied others in post-coloniality.* London: Routledge.

Blackman, L. (2004). Self-help, media cultures and the production of female psychopathology. *European Journal of Cultural Studies.* 7(2), pp. 219-36.

Brancusi, C. (1907-8). The Kiss. In *The Essence of Things*. Tate Modern, London: 29 January-23 May, 2004.

Callon, M. and Law, J. (2004). Introduction: absence-presence, circulation, and encountering in complex space. *Environment and Planning D.* 22(1), pp. 3-12.

Cohen, A.P. (1987). *Whalsay: symbol, segment and boundary in a Shetland island community.* Manchester: Manchester University Press.

Davidson, J. and Smith, M. (1999). Wittgenstein and Irigaray: Gender and Philosophy in a Language (Game) of Difference. *Hypatia.* 14(2), pp. 72-96.

Giddens, A. (1992). *The transformation of intimacy.* Cambridge: Polity.

Goffman, E. (1969). *The presentation of self in everyday life.* London: Allen Lane The Penguin Press.

Graham, J.J. (1999). *The Shetland dictionary.* Lerwick: Shetland Times.

Irigaray, L. (2000). *To Be Two.* London: Athlone.

Jackson, P. (1998). Constructions of 'whiteness' in the geographical imagination. *Area-Institute of British Geographers.* 30(2), pp. 99-106.

Jamieson, L. (1998). *Intimacy: personal relationships in modern societies.* Oxford: Polity Press.

Kincaid, J. (1988). *A Small Place.* New York: Farrar, Straus and Giroux.

Kobayashi, A. and Peake, L. (1994). Unnatural discourse. 'Race' and Gender in Geography. *Gender, place and culture.* 1(2), pp. 225-43.

Kristeva, J. (1991). *Strangers to ourselves.* New York; London: Harvester Wheatsheaf.

Massey, D. (1997). A Global Sense of Place. In *Reading Human Geography: The Poetics and Politics of Inquiry.* (Eds, Barnes, T. and Gregory, D.) London: Arnold, pp. 315-23.

Milligan, C. (1999). Without these walls: a geography of mental ill-health in a rural environment. In *Mind and body spaces: geographies of illness, impairment and disability.* (Eds, Butler, R. and Parr, H.) London: Routledge, pp. 221-39.

Morley, D. (2001). Belongings: Place, space and identity in a mediated world. 4(4), pp. 425-49.

Parr, H. (1999). Bodies and psychiatric medicine: interpreting different geographies of mental health. In *Mind and body spaces: geographies of illness, impairment and disability.* (Eds, Butler, R. and Parr, H.) London: Routledge, pp. 181-202.

Parr, H. (2002). Rural Madness: Culture, Society and Space in rural Geographies of Mental Health. AAG: Los Angeles. March.

Probyn, E. (2003). The Spatial Imperative of Subjectivity. In *Handbook of cultural geography*. (Ed, Anderson, K.) London: Sage 2003, pp. 290-9.

Robertson, T.A. and Graham, J.J. (1991). *Grammar and usage of the Shetland dialect.* Lerwick: Shetland Times.

Rose, N. (1990). *Governing the soul: the shaping of the private self.* London: Routledge.

Sedgwick, E.K. (2003). *Touching feeling: affect, pedagogy, performativity.* Durham, [N.C.]; London: Duke University Press.

Thien, D. (2005). *Intimate Distances: Geographies of Gender and Emotion in Shetland.* Unpublished PhD thesis, the University of Edinburgh.

Wittgenstein, L. (1953). *Philosophical investigations.* New York: Macmillan.

Chapter 15

An Ecology of Emotion, Memory, Self and Landscape

Owain Jones

[T]he faltering and fading sounds which I think lingered on in me at least for a while, like something shut up and scratching or knocking, something which, out of fear, stops its noise and falls silent whenever one tries to listen to it (Sebald, 2001: 195).

The levels hiss and hum in the warm spring sunshine, faintly ooze and crackle with drainage. 'Ok!' – one of my bigger brothers. A battered iron gate scrapes open – 'hoowww, get on!', a man's voice, loud, rich, slightly Welsh – my dad – he whistles for the sheepdog. The flock of sheep, previously pressed against the gate, begin to trickle over the small stone-arched bridge and onto the coast road, scuffling and bleating as they go.

Memories mobilise, a landscape within me comes alive, reforms, yet into something fresh. I change.

Introduction

My intention in this chapter is to offer an exploration of emotion, memory, self and landscape. These are, of course, deep territories, and when conjoined, as I think they inevitably are, their complexity is exponentially compounded. So what follows is a mere sketch of a small moment/process/locality in the vast universe of a self, the purpose of which is to highlight the importance of memory within geography and emotional geography.

Life is inherently spatial, and inherently emotional. As Damasio (2003: 4) points out:

feelings of pain or pleasure or some quality in between are the bed-rock of our mind. We often fail to notice this simple reality [ ] But there they are, feelings of myriad emotions and related states, the continuous musical line of our minds.

Emotions are systemic and interact constantly with our conscious and unconscious selves, memories and environment; they enframe the rational and not vice versa. So who we are and what we do at any moment is a production of the stunningly complex interplay between these processes. These emotional spatialities of becoming, the transactions of body(ies), space(s), mind(s), feeling(s) in the unfolding of life-in-the-now, are the very stuff of life we should be concerned with when trying to understand how people make sense of/practice the world. The emotional associations of memories are a key dimension to be considered within these processes and therefore within emotional geography and geography more broadly.

Each spatialized, felt, moment or sequence of the now-being-laid-down is, (more or less), mapped into our bodies and minds to become a vast store of past geographies which shape who we are and the ongoing process of life. The becoming-of-the-now is not distinct from this vast volume of experience, it emerges from it, and is coloured by it, in ways we know and ways we don't know. If we are all vast repositories of past emotional-spatial experiences then the spatiality of humanness becomes even deeper in extent and significance.

But Harrison (2002: 3) has suggested:

> There is something about emotions, or about 'emotional experience', that troubles the operation of social theory; that resists being bought into the thematisation, conceptualisation and systemisation that must be part of any social analysis. This is not a grand claim [ ] It is simply to say suggest that there is something about 'emotional experiences' that eludes our attempts at recollection, which resist representation.

Can we recollect past emotion-spatial experiences for the purpose of some attempt at representation? Can we go back to the past terrains and past encounters which are mapped inside us and which colour our present in ways we cannot easily feel or say? One way of trying to do this is to turn to 'other' forms of writing, to devote our efforts to description and narrative rather than the treadmill of theory, evidence and analysis. The extracts (italics) are an attempt to write more fully an account of memories of a childhood landscape.

> *We are leaving the furthest point of the territory of our farm through the 'Iron Gate' which opens onto the coast road. We are on the moors, the Wentlooge Level, the strip of nearly forgotten coastal land that runs between the docks of Newport and Cardiff.*
>
> *My father drives behind the flock in our car. A horse box in tow, carrying the odd lame ewe, a few fence posts and tools, stuff for dosing any sick looking stragglers, a bag of feed, and hand sheers for clipping wool. The knife for paring their overlong hooves, sometimes cutting out the rot, is in his pocket. He toots the horn, hustles the sheep – and us. From in the car, window open 'geeertt oooonnn, sheeeep! All right Nimrod?'*
>
> *A bit impatient, 'come on Narbus!'*
>
> *I am 'Nimrod' nearly seven, my sister, a little older and a little bigger is 'Narbus'. My dad is a nick-name genius.*
>
> *We walk by the car. We can ride if we get tired.*
>
> *We are driving sheep west toward the Lamby, an outpost of seasonal tidal grazing land in the heart of Cardiff. A small wilderness on the moors by the docks, cut off by river, seawall and railway, used by us and a few other farms. One of my big brothers walks in front of the sheep, a vanguard with calls – he sings out up ahead, 'Cooome ooonnn!' – and a stick. We all have sticks if we want them, straight hazel coppice, lying along the floor in the back the car, cut from 'the special place' in the hedge back home.*

Another even bigger brother sits on the front of the bonnet of the car as its tyres roll stickily along the quite, flat, straight road. The dog Pedro weaves behind the sheep, tongue lolling and dripping in the increasing warmth, harassing any who dare lag from the tight pelaton they form.

A field away to the left the sea wall, a regular earth bank which holds back the high tides from the low levels, follows our direction. We will follow it west to Cardiff, to where it ends. Inland, to the right, a few more fields away, the London to South Wales railway, on an embankment like the sea wall, also heads for Cardiff, curving across the green reen-gridded moors.

'I want to get in!' 'And me', says my sister. 'Come on, come on.' The car stops and starts to roll again before the doors are closed.

My dad seems to wear the farm, or merge with it. His cloths are checked, brown, corrugated and sour. Pockets bulging with baler twine, packets of stuff, nails, all mixed with bits of straw, and always the knife.

As we crawl slowly along behind the sheep, he jiggles his hands on the wheel and the gear stick, talking excitedly under his breath to the car, like he did to the horses he once worked with (so my brothers said) before my time, jumping his shoulders in time. He's mumbling quickly, under his breath, almost unintelligible, 'get up, get up there, come on, come ...' then loud, 'how! Geerrtt oooon sheeeep!'

Emotional Geography, Memory and (Geographical) Imagination

Developing the notion of emotional geographies, Anderson and Smith (2001: 7) write that 'to neglect the emotions is to exclude a key set of relations through which lives are lived and societies made'. Soon after this first, vital, principle comes the realisation that emotions are intensely political, gendered, and spatially articulated in many obvious and less obvious ways. This call to heed emotion can be seen as part of the movement away from the claim that knowledge is, and should be, an abstract, disembodied, purely rational and objective construct. It recognises the role of emotions in the *construction* of the world, and in *interpretations* of the world. It continues the movement to accept the *full humanness and complexity of being-in-the world*, clearly linking with the work of geographers who are trying to confront non-cognitive knowledges of the unconscious, the body and so on (Thrift, 2001). A basic element of Thrift's approach is that as much of the processes and forces of 'social' life lie beyond the realm of rational thought and language, to remain in that realm in terms of focus and method is to miss out on, and misunderstand, much of what life is. One clear implication of this is that the emotional dynamics of the social and self need consideration. Not only do the vast stores of emotional-spatial experience embedded in us influence our (spatialized) practices of the now, they will also shape, as geographers, in combination with many other influences, our substantive, theoretical and methodological predilections.

The road has wide lush verges which tempt the sheep to linger, despite the dog's attentions. Outside these, two reens, the ditches which drain the levels, follow the road. Each field we pass on either side, also defined by reens, has a small ramshackle bridge crossing to it. Some are old stone arches, others newer, made of concrete and railway sleepers. More substantial ones cross to the farms that we pass. Maerdy Farm, Sluice House Farm, Ton-y-pill Farm, Sea Bank Farm, Swn-y-mor Farm; my father knows the owners, the families and the fate of everyone. Some of them are relations. He passes comments, 'old Billo Harris's cattle are looking well ...'. 'I want to get out' my sister says. 'Me too.'

Our bigger brother has swivelled on the bonnet and slid off the side of the car straight into a walk. He follows behind the sheep, swishing his stick in sweeping sporting strokes. We follow. He is nearly as tall as my father, but not as big. Shirt tied around his waist, tousled black hair against the sky.

Me and my sister, small figures, without duties, drop behind and explore briefly. Throwing stones through the perfect speckled green skin of duck weed on the reen, or trying to sink the occasional floating bottle which, when hit, disappears with satisfying completeness. Then we run to catch up.

The road turns north with the coast. It then rises, and we cross the railway. The moors, the approaching city, the mysterious inland hills, the curling trajectory of the line can all be seen, and then we descend again to reen level. The billowy textures of willow and long grass, and the scope of the sky make a stuttering change to suburban. The sounds are harder, the green is gone, the sky is crudely cropped.

Memory and Space

Memories and emotions – in all their forms – are the ecology on which the 'mega-fauna' of our consciousness, our rationality, our academic thoughts depend. We are creatures of memory. This is so in the very real sense of the partial collapse of self of those who have severe memory damage. Memory along with the unconsciousness makes being a self possible, it weaves the consciousness and unconscious together. 'The self – the identity – is always in flux, the present altering the past even as the past informs the present' (McConkey, 1996: 315).

I want to briefly develop two points; firstly about the complexity and uncertainty of memory. Damassio (1994) shows that we don't control memory, we are not aware of it working in the same way as conscious reflexivity. We are not aware of, or in control of, how experiences are mapped into us at the moment of their living out, or of how they are retained and retrieved (or not) through differing forms of memory. (The same goes for emotions; we can be reflexive about our emotions but emotions are not themselves reflexive). Also memory is not just a retrieval from the past or of the past, it is always a fresh, new creation where memories are retrieved into the conscious realm and something new is created. The strangeness of memory is the presence of what is apparently past in the present.

Secondly, I want to consider the spatial nature of memory and how some geographers (and others) are addressing this, for these hint at the significance and complexity of the ideas of the geographies of memories and memories of geographies and how we might think about them. Philo's (2003) paper on childhood turns to Bachelard's notion of reverie, an episode of memory when we somehow travel back. A state where consciousness can slip back towards a more dreamlike state, where the imagination, freed from the firm direction of focused thought and action, can begin to 'drift' back into all the remembered spaces, events and feelings which are in our minds. Is this a means, Philo asks, of getting back 'into' childhood, can we relive past emotional-spatial experiences?

Game (1991, 2001) is interested in another form of memory process, namely, involuntary memories. Different in quality from reverie, these are sudden sharp moments when the impression of some past place or event springs fully formed into one's mind. Are these the 'random noise' of the memory process or are they markers of the significance of both past moments and the present moments which somehow may summon them?

A number of people have turned their attention to ideas of body memory. Casey (1987) develops a whole typology of body memories including habitual, traumatic and erotic. Weiss (1999: 33) suggests that:

> previous body images [which] remain accessible and can be re-enacted in a moment as when we return to a childhood 'haunt' and find ourselves simultaneously haunted by an earlier body image that was able to negotiate the childhood space with ease. These earlier body images are also projected onto our own children as we watch their fascination with/dread of their own bodies as we find ourselves inhabiting their ways of living their bodies as the emotional center of the world.

Much has been written, in numerous contexts, about the possibility, or otherwise, of truly recalling past experience, and of the adult relationship with past child self. Do we remember chiefly through narrative? Can we remember emotional experience? To what extent are our emotions of the present driven by memory in obscure(d) ways? These ideas, and the complexities and challenges they raise, serve only to complicate and confound any simple path to a clear conclusion. But as humans are oblique creatures of memory and emotion this is only to be expected.

Mantel (2003) brings an interesting spatial analogy to understanding memory. She feels the archaeological notion of memory, with the strata of distant past being below the strata of the most recent memories might be misleading, and perhaps a better analogy of memory is some vast space – 'St Augustine's "spreading limitless room" or a great plain, a steppe, where all the memories are laid side by side, at the same depth, like seeds under the soil' (p. 25).

For Probyn, memory, and writing about memory is a spatial rearrangement by which 'the past is bent into strange shapes so that what should be furthest away is in fact the closest' (1996: 113). She further adds that 'images of childhood, from childhood, pull us back to a space that cannot be revisited; they throw us into a present becoming, profoundly disturbing any chronological ordering of life and being' (p. 103).

> *Our drove has entered a knuckle of suburb – Rumney – a 1950s estate jutting out from the fist of central Cardiff.*
>
> *Now we do have a duty. We have to run ahead of the sheep and stand to stop them turning off onto other roads or into tempting front gardens.*

'Howww! get on!' shouts my dad from the car. 'All right you two?' 'Go on then!'. And once the flock has past us we run on again to overtake and man the next gap or junction.

We ran lightly and tirelessly, as only children can, and it felt like we could run forever. We could not sense the inevitable limit – not of our happiness, exactly – but of our being in this place and time, and of the place itself.

People come out of their houses to watch, some to fuss over their gardens which, once in a while, are invaded with maximum drama. A sheep casually takes some gleaming flower hanging through a low fence.
    Other children follow, shouting and running too. But this is our glory.
    Then, suddenly, we turn left, through a break in a long terrace of houses and shops. The tarmac stops, and the track is gravel. The scruffy back fences turn back into field hedges, and then we rise up a short incline and cross our bridge (a farm crossing), back over the railway towards the sea, from the tatty suburb back into a cul-de-sac of countryside. From the height of the bridge it can all be seen, the Lamby, the flat reaches of the moors stretching down to the shore, the industry and docks of Cardiff to the west.

Memory, Imagination and Emotion

Memories always will have a spatial frame (even if it is unremembered or latent) and they will be always be emotionally coloured in hues ranging from pale to vivid. One way to think of emotional geography is to think of the connections between memory and our geographical imaginations. Memory must play a key, formative role in the construction of our ongoing emotional and imaginative geographies.

> the 'key elements in an individual's autobiography', as well as her or his primordial and recent body states, are represented within the brain – all of them constituting representations crucial to emotion (McConkey, 1996: 60-61, citing Damasio, 1994).

The relationship between memory and imagination that Philo (2003) discusses in relation to children's geography seems important not only in that context but in the wider context of the much discussed 'geographical imagination'. As Warnock (1987: 76) writes 'we could say that in recalling something, we are employing imagination and that in imagining something, exploring it imaginatively, we use memory. There can be no sharp distinction'. So I want to point to these strange geographies which occupy us all, which hover between the then and the now, between our geographical imaginations and our geographical memories, to these hybrid ecologies of the self and to the other element, their emotional register.

Memories, according to Hampl (1995) consist of image and feeling, the event and the response to that event. This distinction is reflexively worked by people using memory as a form of 'research'. Toni Morrison has stressed that when remembering and remembering to feed her writing, she first and foremost responds to the emotions of some past encounter rather than the narrative of that encounter (cited in McConkey, 1996: 212). So we have the possibility of remembering reflexively, the narrative of an event, or somehow remembering in different ways, and perhaps remembering emotionally when re-feeling past feelings. Hampl says she does not want to remember through 'the grating wheels and chugging engine of logic' but instead through the heart – 'the guardian of intuition with its secret, often fearful intentions [whose] commands are what the writer obeys' (Hampl 1995: 206).

A train horn sounds the so familiar two-tone note. A goods train passes under the bridge, the slow dud-dud – dud-dud, dud-dud – dud-dud rhythm softly shaking the immediate world. We pivot on our tummies and watch it curve away through the industrial hinterland of Cardiff, maybe spit into the coal wagons. The outlines of colliery spoil heaps are visible on the distant, hazy horizon. Here, the sea wall turns inland, and, nearly meeting the railway which is still on its embankment, it leaves 150 acres of tidal grassland open to the higher tides of the estuary. The third margin of the Lamby is made by the Rhymney river flowing down from the inland hills, from up 'the valleys'. It wanders with a few final meanders to the sea, and the Lamby is held in the eastern folds of these last meanders, isolating it from the city.

The gate at the bottom of the right-angled ramp on the other side of the bridge is held slightly ajar by my father, the flock pressed against it by my yelling brothers, and the animals nearest the slightly open end of the gate, sensing escape, suddenly spring forward and out onto the Lamby. My father counts them, keeping up with the flow of the animals, which could surge suddenly if their jostling, collective weight forced the gate wider open against his own considerable weight and strength.

We tried to count, but we never could keep up. Some of the sheep made impressive bounds as they escaped. Then they trailed away along thin brown tracks worn into the stiff grass by previous generations of stock brought here for summer grazing, and onto and over the seawall and down onto the saltings, the tidal wharfs.

From then, until the sheep, or sometimes it was cattle, (the droving of which was quite another, more erratic and rowdy matter), were counted again and driven back through the suburbs, back along the coast road again in the autumn to the sanctuary of the home fields and the barns, Sunday morning visits to the Lamby to check on the stock were a feature of our lives. We left the house, the yards, and the close at hand fields to go 'down the Lamby'.

While my dad and brothers looked over the animals, my sister and I were free to wander and explore. To the west the city rumbled quietly on the low again horizon. Trains thudded past in the distance. The space was open, wide, flat, a continuation of the line of the south horizon out on the estuary. A huge sky started out there, below eye level it seemed, and domed hugely over and around us until it fell onto the inland hills to our backs. The docks darkly frayed its bottom edge where the sea invisibly met the land just down the coast.

I remember the reaches of wiry grass stretching far away from our feet, patterning silver in the wind which, coming off its sea routes, usually had something to say. Sometimes it blew hard, pressing tears from our eyes, sometimes warm, sometimes cold, sometimes wet, but it always carried the dirty, salty freshness of the estuary. And it carried the distant, ragged cries of gulls that reminded us that this was indeed the sea. Often its entire margin was indistinct out in the vast surfaces of glinting mud. Sometimes it was a grey roughness marching on, or retreating from the shore, depending on the tides.

But above all it was the lonely, scurling cries of black headed gulls which marked this place and echoed its lonely wildness and, which, when I hear them now, prompt a thin creeping tide within me, in my mood and in my body, but a tide of what I cannot easily say.

The grass was lined by thin meandering braids of jetsam left by the mid range tides. Sometimes three or four in broken, rough parallels, marking the retreating extent of days of lessening tides. Ascending tides would pick up these previous lines and make them into one heavier fresh one, to be moved on yet again at the next high tide.

Sometimes it was all still, except for the heat simmering the horizon, the close around but distant city rumbling softly, the trains passing, and the black-headed gulls calling as they moved down the river to the shore.

Tiny streams began in the grass and snaked down to the shore, growing into miniature canyons in the mud, a challenge to jump, or follow down to the ragged edge of the land. This was a zigzaggy mud cliff, about six feet high with the wiry turf making a quiff over-hanging the top. This marked where the levels, gave way to the flowing, oozing mud beds of the estuary. Small islands of green, dying fragments of turf, standing atop crumbling columns of soil split away from the cliff, tilting and sinking at crazy angles, studded the near of the reaches of mud that glistened and veined out to the never the same distance away water line.

Turning around, the bridge, the car on the sea wall, our father and brothers amongst the stock, were tiny, far away, almost hidden by the flatness.

On the other side of the Rhymney there were allotments, with sheds made of scrap wood, old doors, even an old rowing boat stood on end. Then there was a modern, concrete motel-cum-pub, with fading pink and cream awnings and weeds flourishing in the always empty car park. And behind them the East Moors steelworks, a series of huge, connected metal sheds, hanging from the sky by fat billowing plumes of white steam, which, every now and then, emitted huge, softly loud, languorous clangs which echoed all around the inland hills.

In the low tide mud of the Rhymney, disreputable old boats stuck at odd angles, painted in garish reds and blues, and dirty whites, the umpteenth layer. Blackened, rickety jetties stuck out from the far bank, constructed and used by some mysterious class of poor urban boatmen.

Undercut by the river on our side, big slithers of land would sheer away and then imperceptibly slide, week by week, down into the water, to make more mud in the endless mud of the estuary. We jumped down onto them, trying to feel their creeping, tilting dissent. But my dad was angry, again, at the river authority and complaining at this erosion of land and property. And if he is nearby, he is impatient, 'come up from there', he chides. And still the trains beat by, the sound more distilled, slowing for Cardiff station, or diesels leaving town, putting on power with a shudder.

We made other, special trips, when the tides were high enough to cover the unprotected land and the stock had to be guided to the seawall in case they got cut-off. At home, in all the chaos of the old desk in the corner of the living room that seemed to bulge with papers and folders like stuff coming out of an old mattress, and which caused rows, the little yellow 'Arrows Tide Timetable' for the Bristol Channel was always in its place.

We stand on the grass in our wellies. The eastern Atlantic is swelling up into a huge shallow dome and moving our way as the moon swings over the earth's shoulder and lines up, for a while, with the sun. The rising water washes along the

South Wales coast, pushes a spearhead of floodwater up the Severn Estuary and deep into the flank of England. It gradually covers all the inter-tidal lands like the Lamby, and then gently starts to examine the flood defences formed of the snaking earth bank walls.

These are the second highest tides in the world. And here, on the Lamby, we played at the precise margin of their progress. You could see the edge of the advancing water creeping through the grass. Then it floated the grass stems, and they swayed with the movement, then it just hid them, and then the feet of our wellies in the silky, heavy brown of the coming flood.

We turn and walk with the tide towards the sea wall, up which we climb and turn to see how high it follows.

At the highest tides the water filled up everywhere, right to the top of the sea wall, leaning on it with what seemed to be a heavy curiosity and longing. For now the Lamby was seabed. We stand on the seawall, the water lapping and running a few inches below our toes. The land to our backs, a thin strip between the sea wall and the railway, is 12 feet below, and here the sheep restlessly huddle.

On the tide ships are moving in and out of the docks, returning, or setting sail onto the great trade routes of the world. A deep long blast of a ship's horn rolls across the flatness and over us like some flexing bubble of plasma, up into the hills, where it stops and rolls back with an echo, and then away over the levels towards home where it might just be felt.

Then, as surely as it came, the water stops rising, pauses, and goes. The gravity which pulled it out over the land is fading as the moon moves on. A blackened, sea-washed tree trunk has been dumped by the tide. It stands out in the distance. We go to see, through a labyrinth of shallow tide pools which mark the lower wharf, and huge swirls of sodden debris left here and there. The whole landscape softly crackles as the last of the water creeps back to its home. The sheep wander back onto the salty grass. We go home.

Memory, Emotion and Place

The child sees everything big and beautiful. The reverie towards childhood returns us to the beauty of the first images. [ ] Can the world be as beautiful now? Our adherence to the original beauty was so strong that if our reverie carries us back to our dearest memories, the present world is completely colourless (Bachelard, 1971: 101).

If as suggested memory is spatial, it is also clearly bound up with processes of place and emotional attachments to place. Casey (1987) argues:

Only consider how often a memory is either of a place itself (e.g. one's childhood home) or of an event or person *in* a place: and conversely, how unusual it is to remember a placeless person or an event *not* stationed in some specific locale (p. 183).

Clearly remembering being-in-place, and perhaps remembering through place, through emotions of (remembered) place are powerful elements of emotional geographies of the self.

For narratives of the self, [ ] are more than just 'situated' in the sense of having a particular, unique time and place. I suggest that narratives of the self are inherently spatial: that they are spatially constituted. Stories of the self are 'produced' out of the spatialities that seemingly only provide the backdrop for those stories or selves (Pile, 2002: 111-112).

Drabble (1984: 7) has noted in her exploration of the relationship between English literature and place, that many writers reveal 'a passionate attachment to the places of childhood' and most 'return again and again to childhood, seeing in a pond, a field, a tree, a church some reminder *of what they once were*' (p. 8). And Paulin (1999: 137) further adds:

I believe there are primal, original landscapes of the imagination. Often they are the places in which we grew up and which remain important to us throughout adulthood [and going back is] like a time warp, a way to hold onto childhood.

But this begs the question of can we go back?

Standing in the middle of the Lamby and looking inland, thousands of tiny windows stare d back from the suburbs of Cardiff on the inland hills. We, the land, the animals, the tides, the birds were part of their view. Other people used the Lamby, some walked the sea walls with their dogs, with sticks and coats, others with binoculars to look at birds and ships. We would inspect the sites of fires made from driftwood, and the tracks made by cycles and scramblers up and over the seawall and the tracks of horses. Unknown others used this wild place right in the guts of the city. In the winter, empty of stock, the hardy users shared it with the gulls, the tides, the wind, the rain and the trains.

At home, there are rows. About 'papers', money, and other things. And discussions about land and ownership. To these, at some point, are added dark phone calls, rages, with the name of the Lamby popping out of them. The names of solicitors and local council officers, linked with curses, become like mantras. In the house, on the phone, shouting, my dad is angry, 'now you listen to me, I'm telling you ...'
Our mother hustles us away, closes the door and tells us not to worry. It means nothing but domestic anxiety to us. To the city council the Lamby now means something else, they want the Lamby as a site for a new rubbish tip. The ever-growing levels of rubbish from the ever-growing city need to go somewhere. And they get their way, in the end, using a CPO (Compulsory Purchase Order) – or the threat of it – I am unclear of the detail. They buy the Lamby. They take the land bit by bit. It was agreed we could still use the rest of it until it was needed. So we still went there. So we saw its slow death.

The sound of the trains was joined by the heavy machinery crawling as the tip slowly grew. Neatly, they used the sea wall as a retaining wall for the tip. It then spread over the flat land down towards the sea, like a reverse, malevolent tide which never intends to retreat. The gulls still cried and scurled but over different fare now. The wind still blew, but it carried that sweet, disturbing smell of rottenness and bits of rubbish. They put up a tall chain link fence to catch it, but the scrabby thorn hedges on the inland side of the tip which marked the few fields before the houses, streamed and flickered with tatty flags of torn plastic. Rubbish was dumped by the gates of the tip at night and slowly scattered around and into the reens. The new road which served the tip, and the new culverts, stood on rough, unfinished concrete fringes on the churned up soil. It seemed to me like a war zone, all previous order gone, normal protocols suspended. All day, bin lorries came, tipped their loads and then went away to get more.

Poisons leaked from the tip and into the reens behind the sea wall, and my dad, fearing for the last generations of stock to graze there, had more tempestuous phone calls and meetings. They straightened the final course of the Rhymney, leaving its last meanders cut off to fester, a new road cut along side it to a new bridge over to where the allotments had been. This all took years. Once I went back to take some photographs. Just past the tip, on the new road, in the middle of the litter, the rubble and devastated land, they put a gypsy park. I took some photos, and they posed for me; old men who looked away from the camera, young men in tight out of style dirty suit jackets, grimy kids, and laughing, devastating teenage girls. Their ponies were the last animals to eat the wiry Lamby grass.

I still have the black and white prints to show that this wasn't some dream. I know our bridge is long gone. It's all gone. A high-fenced, golf-course green hill now reaches down to where the soft shore had been. A swelling like an ill, bloated belly, with a protruding navel of shiny tanks and valves, venting methane and the last ghostly airs of the flat wild land that once was there. The hill violates the flatness of the landscape, hides the estuary, confines the scope of the sky. Starter industrial units built by the Welsh Development Agency stand around it in the raw landscape, mostly empty. The river is there – somewhere, in a concrete channel.

'Topocide: the annihilation of place' (Porteous 1988), was a geographical concept I soon seized upon in my first postgraduate studies. Porteous argues that 'topicide is an emotional issue (p. 78) and one which has rarely attracted geographers' (p. 75).

Conclusion

I have tried to *do* rather than just *say* in this chapter, I have tried to articulate some of the spatialities of my memories through description and narrative. Why I chose them, or they me, I cannot easily say. Geographies of memories and memories of geographies are complex and emotional, not least because some, perhaps the most powerful, will be the geographies of our childhood.

Within his analysis of images of *The Town and the Country*, Williams (1985) considers Clare's lamentations for the loss of the countryside of his childhood in the enclosures and other agricultural developments of the 18th century. Williams extracts from the Clare's poems that the 'primitive land' was 'being directly altered: the brooks diverted, the willows felled, in drainage and clearance' (p. 138). Williams adds that 'the particular trees, and a particular brook, by which I played as a child, has gone in just that way, in the last few years, in an improved use of marginal land' (p. 138), and he goes on to explore how the landscapes that Clare laments were the landscapes of his childhood, *and it becomes uncertain which is being mourned.* The emotions of childhood are imprinted onto whatever landscape they are acted out in.

Figure 15.1 Me with a bigger brother on the farm

Source: Photograph owned by author

Figure 15.2 The scene of where the farm once stood

Source: Photograph owned by author

Figure 15.1 is picture is of me with one of my older brothers on the farm we returned to after our trips down the Lamby. This is the farm where I grew up. This too, like the Lamby, has all gone, the farm, yard and barns beneath the estates of East Cardiff built in the late 1970s (Figure 15.2). The house was turned into flats.

So it has all gone, doubly – as time and as space. I get a feeling of panic, that my whole existence is thinned as the spaces of the past have been eradicated. They are mapped into my memory, re-form in my dreams, and form hybrid landscapes with other places I have known or know now.

But this is a story of losses of kinds which many have faced, and in a way, which we all face, the loss of past geographical selves. It is difficult because there is so much that could be said and could be shown.

When searching through the photos I have of the farm and family life there, I found myself leaning right forward, staring at details of the pictures as they came up on the pc screen as if half hoping it was all going to come alive again. I feel overwhelmed by each photograph. Why am I so moved? Why am I so concerned? Inevitably this is lived and personal, the divisions between the public and professional and the private are breeched. This is the point and the challenge of emotional geographies.

References

Anderson, K. and Smith, S.J. (2001) Editorial: Emotional Geographies, *Trans Inst Br Geog*, NS 26. 7-10.

Bachelard, G. (1971) *The Poetics of Reverie: Childhood, Language and the Cosmos*, Boston: Beacon Press.

Casey, E. (1987) *Remembering: A Phenomenological Study*, Bloomington: Indiana University Press.

Dasmasio, A. (2003) Mind Over Matter, *Review, Guardian*, 10.05.03, pp. 5-7.

--- (1999) *The Feeling of What Happens: Body Emotion and the Making of Consciousness*, London: William Heinemann.

--- (1994) Descartes' Error and the Future of Human Life, *Scientific American*, vol. 271, Oct, p. 105.

Drabble, M. (1984) *A Writer's Britain: Landscape in Literature*, London: Thames and Hudson.

Game, A. (2001) Belonging: experience in sacred time and space, in J. May and N. Thrift (eds) *Timespace: Geographies of Temporality*, London: Routledge.

--- (1991) *Undoing the Social: Towards a Deconstructive Sociology*, Buckingham: Open University Press.

Hampl, P. (1985) Memory and Imagination, in Hunt, D. (ed.) *The Dolphin Reader*, Boston, MA, Houghton, 1003-1014.

Harrison, P. (2002) 'How Shall I say It ...?' Emotions, Exposure and Compassion, paper given to Emotional Geographies Conference, Lancaster University 23–25 Sept 2002.

Mantel, H. (2003) *Giving Up the Ghost: A Memoir*, London: Fourth Estate.

McConkey, J. (ed.) (1996) *The Anatomy of Memory: an Anthology*, Oxford: Oxford University Press.

Paulin, T. (1999) Tom Paulin, in S. Smith (ed.) *My Country Childhood*, London: Coronet Books.

Philo, C. (2003) 'To Go Back up the Side Hill': Memories, Imagination and Reveries of Childhood, *Children's Geographies*, 1, 1, 7-24.

Pile, S. (2002) Memory and The City, in J. Campbell and J. Harbord (eds) *Temporalities, Autobiography and Everyday Life*, Manchester, Manchester University Press.

Porteous, J.D. (1988) Topocide: The Annihilation of Place in J. Eyles and D.M. Smith (eds), *Qualitative Methods in Human Geography*, Cambridge: Polity Press, 75-93.

Probyn, E. (1996) *Outside Belongings*, London: Routledge.

Sebald, W.G. (2002) *Austerlitz*, London: Penguin Books.

Thrift, N. (2001) Still Life in Nearly Present Time, in P. Macnaghten and J. Urry (eds) *Bodies of Nature*, London: Sage.

Warnock, M. (1987) *Memory*, London: Faber and Faber.

Weiss, G. (1999) *Body Images: Embodiment as Intercorporality*, London: Routledge.

Williams, R. (1985) *The Country and the City*, London: The Hogarth Press.

Chapter 16

On 'Being' Moved by Nature: Geography, Emotion and Environmental Ethics

Mick Smith

Introduction

What might it mean to take the idea of an emotional and ethical geo-graphy seriously? Could our ethical feelings towards the Earth and its inhabitants be (re)inscribed in a language of closeness or distance, bounded or open-ness, space and/or place that might express something of our environmental concerns? Could talk of being 'moved' by nature be regarded as more than merely metaphorical? If so, then geography might cease to be understood merely as a disciplinary field or practice to which pre-existing ethical theories are applied as though they were sticking plasters. Rather, we might begin to explore the myriad possibilities for writing and valuing a world, the affective and ethical significance of which has been marginalized by modernist geographical and philosophical paradigms. Perhaps certain geographies might become expressions of environmental ethics, of our passionate concerns for 'natural' places.

Of course, to speak of nature and environment in such ways is already to risk falling into a taxonomy which radically separates these things from us, classifying them as the not-human, as that which surrounds us externally. Such distinctions, which regard language and ethics as purely human characteristics and feelings as something entirely internal and subjective make even broaching topics of environmental ethics or emotional geographies so very difficult. This chapter must begin then by trying to show how ethics and emotion are modes of attachment to and involvement in a world that is never just composed of abstractly defined relations to our fellow (human) citizens, still less confined to some isolated ego. Rather, emotion and ethics are key features of those intimate participatory practices that draw us closer to others, affecting our modes of being-in the-world, giving us a feeling for and an understanding of our relational emplacement within that world (see also Conradson, this volume). To speak of how this understanding might be theorised as well as felt, I'll then deploy the work of two philosophers more usually associated with very human-centred views of language, Hans-Georg Gadamer and Ludwig Wittgenstein. Both, I'll argue, can help us develop an understanding of how ethics and emotion are part of producing a *meaningful* world, a world worth caring about.

We might begin then by asking why geography and ethics are currently deemed so distant from each other that they are rarely conjoined in anything but the most superficial manner. The ethical paradigms that have dominated modernity, like utilitarianism and rights, have certainly had geographic pretensions, but only insofar as in laying claim to a universal writ they effectively deny the social and historical particularity of their own origins. As Bauman (1993: 39) argues moral universalization requires that we recognize 'as moral only such rules as pass the test of some universal, extemporal and extraterritorial principles'. Modernist morality subsumes geographic particularity under its global ambitions with the corollaries that only that which is universalizable counts as ethical and that which is ethical must be universally applied. These universal standards are insulated from the dangers of historical or geographic relativism and from moral subjectivism because they are supposedly dependant only on reason's uncanny ability to establish ethically neutral arguments, procedures, or codes, that can then form the basis for consensus, or at least negotiation, between disparate positions. Reason thereby places a limit on the vagaries of dangerous and unpredictable passions. Our feelings for others, whether human or non-human, are either excluded altogether or enter the moral equation only indirectly in terms of a calculus of self-interested pleasure and pain.

The necessary cost of this rational universalization then has been both the geographical impoverishment of ethics and the exclusion of the personal and passionate engagement with others that has always been the font of our ethical concerns. But as Bauman argues, 'to delegitimize or "bracket away" moral impulses and emotions, and then try to reconstruct the edifices of ethics out of arguments carefully cleansed of emotional undertones and set free from all bonds with unprocessed human intimacy, is equivalent ... to saying that if only we could get the walls out of the way we would better see what supports the ceiling. It is the primal and primary "brute fact" of moral impulse, moral responsibility, moral intimacy that supplies the stuff from which the morality of human cohabitation is made' (Bauman, 1993: 35).

A post-modern environmental ethics, critical of any emotionally cleansed abstract universalism might be thought of in terms of reconnecting this 'brute fact' about the roles of passions and persons with our experiences of co-habiting with non-humans and with our understandings of particular places (contexts and environments). It might argue that an ethical relation to our environment does indeed entail a manner of our 'being-there' (*Dasein*) that finds itself emotionally re-oriented in relation to significant (and often non-human) others. Any geography associated with such a view then becomes an attempt to understand, express, and *interpret* these ethical relations spatially, a hermeneutics of moral dimensions that tries to explicate how ethical feeling and action emerge from a fuller understanding of *where* we come from and find ourselves. This is not a matter of fixing or mapping the contours of our present existence but of understanding our environmental entanglements and how these might recompose our mode of being-in-the-world. The process of ethical becoming requires an emotional openness to circumstance that enables the previously determined boundaries of our being to be re-constituted and re-interpreted. It thus shifts the grounds of our being in terms of emotion and understanding and re-situates us within an ethical relation experienced as a felt need to *conserve* a space appropriate for the continued existence and expression of others, human and non-human.

The places and contexts in which these ethical geographies might emerge will clearly differ from culture to culture, person to person, across space and time. This is not to say they are 'subjective in' the sense of being incommunicable to others; far from it, it is merely to recognize the full complexity of the social and environmental forces that are the font of our feelings. Social and natural history combine in often unpredictable ways to engender emotional attachments and ethical concerns.

Emotional Flows and an Ethics of Place

Near Ilam in the English Peak District, the land of my birth, a river meets itself. In wet weather the River Manifold flows South through a deep limestone gorge. But it also has another more secretive route. Several miles upstream at Darfur Crags it disappears into a swallow hole, and for most of the year when its over-ground flow is diminished, it vanishes altogether from sight until rising again just above Ilam Hall. Here as the river pushes its way back up to the light in a series of springs it emerges as if from nowhere into itself, bubbling, swirling, forcing its way up from unseen passages, pouring forth through rocks and its own now full river bed. This welling up of water within water is a thing of great beauty and mystery. It may well have given Ilam its name – perhaps a corruption of the Welsh for a place of springing waters (Smith, 1995). The river too is aptly named, its many outlets from unknown chambers, its myriad manifestations, the multiplicity of forms 'furnished by sense' (OED) that resist being unified in the synthesis of rational understanding. As you stand and watch clear water appear into clear water there is no point of origin, no obvious channels, no way of predicting what will happen – just the constantly shifting patterns as it flows through and wells up within itself. A mile or so downstream its water will mix again with the waters of the Dove.[1] This place affects me more than I can say.

Why should this be? 'Water', says Bachelard (1999: 14), 'causes springs to gush forth. Water is a substance that we see everywhere springing up and increasing. The spring is an irresistible birth, a *continuous* birth. The unconscious that loves such great images is forever marked by them. They call forth endless reveries.' These dream-like reminders of birth and existence do indeed mark the local culture of the Peak District. In a landscape where waters appear and disappear, local festivities centre on 'well-dressings'. Every summer village wells are adorned with religiously inspired and intricately executed scenes made of flower petals.

But are Bachelard's oneiric musings and the outmoded traditions of a few villages enough to imply that nature can or should play a creative role in our cultural sensibilities and values? How can springs enter my soul, affect my emotions, and imbue me with deep-seated feelings of joy or mystery, desire and wonder? How can these natural events and places become part of the ethical topography of my life, sources and objects of moral sentiment? After all, Luc Ferry argues in his critique of environmentalism, as far as modernity is concerned '[n]ature is a dead letter for us. Literally it no longer speaks to us for we have long ceased – at least since Descartes – to attribute a soul to it or believe it inhabited by occult forces' (Ferry 1995: xvi). If this is so, then being 'moved' by nature must surely be indicative of a pre (or perhaps post) modern irrationalism. It is to find meaning and morality where none exists, to invest the alien 'otherness' of nature with characteristics that are found solely within the remit of human sociality.

This, of course, is also the accepted premise of nearly all 'modernist' theories of language, meaning and ethics. Yet I want to demur. Like many interested in developing an environmental ethics I want to argue that the natural world does *mean* something to me. Springs do in some sense 'speak' to me, they affect me, move me, altering my understanding of my relations to my surrounding environment. Their activities make me *attend* to the modes in which they present and express themselves and, just as with other humans, they thereby acquire meaning and value, they become significant. Contra Ferry, I will argue, nature is a dead-letter only to those 'moderns' who have lost the ability to listen to and interpret the non-human world.

This, I think, requires exploring the inter-relations between three aspects of what might constitute our ethical relations to nature. First, 'natural expression', that is the expressive activity of beings of all kinds demonstrating their continuing existence (being) to others. Second, 'interpretation', that is the attempt to understand, to get a feeling for and give a meaning to what the other expresses (and hence their being). Third, 'affection', that is the development of an emotional and ethical disposition towards others, one that can lead us to sustain them in their difference from us. The dominant tendency in modern philosophy as Ferry illustrates has been anthropogenic and anthropocentric, it claims that all three of these relations are essentially human capabilities. Only human language provides a medium capable of meaningful expression, only human reason a reliable means of interpretation, only a human being is capable of an ethical relation to others (and, moreover, those others are themselves exclusively human). These, then, are the prejudices of our age, the traditional and largely unquestioned assumptions regarding humanity's central place in nature that constitute the contemporary horizons of our understanding.

Such talk of prejudices and the horizons of understanding brings to mind Hans-Georg Gadamer's concept of 'effective history' (*Wirkungsgeschichte*), that is, the 'traditions' that constitute the limits and possibilities of our understanding. These traditions might include those narratives, vocabularies, plots, trajectories and so on, current in our society, the practices and 'ideologies' into which each of us is 'thrown' and through which we try to understand ourselves and the world about us. Modernist philosophical discourses have tended to regard such geographically and historically variable cultural traditions with suspicion, as infecting disinterested rational discourses with relativistic biases. But Gadamer argues that all understanding, even the kind of universalistic comprehension sought by modernism, is actually predicated on and formed through a world of previous understandings developed and made available within specific cultural horizons. For Gadamer then, the term 'prejudice' does not have the usual negative connotations since every attempt to understand involves drawing upon the pre-judgements available to us.

Prejudices are, Gadamer argues, necessary; we both embody and project them onto our surroundings. They have a formative influence, they are the 'effective human historicity' that provides us with a particular cultural 'home' that becomes second nature to us. But neither history nor ourselves stand still and through incessant acts of interpretation we gain new insights, become aware of other different understandings which can challenge and modify the history bequeathed to us. Gadamer's work thus provides a basis for understanding this process of 'becoming' someone different through openness to the differences expressed by others. In emphasizing the importance of historical and social context it also provides a direct challenge to the presuppositions of universal moral reasoning.

Unfortunately, Gadamer never challenged other key aspects of the dominant humanist tradition of his own modernist culture. His work is anthropocentric in all the senses mentioned above. The traditions of which he speaks are the products of human culture expressed and interpreted through human language. We might say that Gadamer's work lacks a natural geographical sensibility. Nevertheless, Gadamer's work might still offer a starting point from which to discuss issues of expression, interpretation and affection in terms of our relations with the non-human world. For not only does Gadamer employ, as I have previously argued (Smith 2001), an *expressive* ontological theory of language, but his interpretative philosophy, his hermeneutics, has important ethical implications. Since language both requires and facilitates communication, a 'sharing in its purest form' (Gadamer 1998a: 7), Gadamer regards it as the original ethical relation. Language 'first raises communality into words' (Gadamer 1998a: 7). The act of interpretation requires one to develop an ethical relation with the text or person one is engaged in interpreting. One has to be alive to differences in meaning to give the other space to express themselves without imposing one's own more limited understanding on them. The task then is to make Gadamer's work provide the basis for a different project, to use his notion of 'effective human historicity' to help develop an understanding of what I'll term an 'affective natural historicity', of the ways the Earth comes to be inscribed in our hearts. In this way we might outline a hermeneutics that could also become a geographically sensitive environmental ethics.

Gadamer and Ethics

Gadamer's (1998a) *magnum opus Truth and Method* begins with an unconventional account of the epistemological, aesthetic and ethical significance of the humanist tradition. Here, we learn that culture (*Bildung*) involves the individual's acquisition of a tacit and tactful understanding of historical and social relations which provides a 'special sensitivity and sensitiveness to situations and how to behave in them, for which knowledge from general principles does not suffice' (16). From Gadamer's perspective there is, one might say, an affective as well as a rational element to understanding one's relations to others. Genuine understanding requires much more that the rote application of formulae, of laws, concepts, methods, or calculus. It needs the individual to cultivate sensitivity to particular situations, to get a *feeling* for their 'place' in relation to history and to others. The general characteristic of *Bildung* is 'keeping oneself open to what is other – to other more universal points of view. It embraces a sense of proportion and distance in relation to itself' (17). The universality that is aspired to here is not some fixed and complete totality, 'a universality of the concept', but the willingness to include the 'viewpoints of possible others' (17). The process of cultivation, of *Bildung*, is one of endless self-formation, of educating oneself in and through openness in one's relations to others. *Bildung* resembles, Gadamer says, the Greek notion of *physis*, (11) the continual upwelling of Nature into the world, though here what is produced is an *ethos*, a practical, but certainly not instrumental, relation to others. This ethos, this feeling for what is appropriate, is quite literally envisaged in terms of developing a common *sense*, a felt experience of understanding how one should relate to others. Gadamer refers to this, as others had before him, as a *sensus communis*, a sense of one's ethical situation in relation to those recognized as significant others.

There is then, for Gadamer, no possibility of an ethical theory in any formal, abstract, or universal sense. There can be no fixed, timeless rules for how one being should relate to another. Ethics is, in Warnke's words, 'ethical pragmatics' (Warnke 2002: 82), a kind of practical reason which, unlike the pure theory so beloved of modernist ethics is not completely divorced from our emotions or from specific material contexts. 'The ethical knowledge of the individual actor ... is the knowledge he or she has of how to act appropriately in a specific situation' and these situations 'vary; they are multifaceted and substantially unique' (Warnke 2002: 82). Self-knowledge, situational knowledge and new experiences thus have to be constantly recombined in a dialectical process of 'becoming' that goes to form our individual ethical characters. 'We make and remake our ethical knowledge and ourselves in those changing circumstances, in the actions we take to apply the ethical knowledge we already possess' (Warnke: 2002: 85). Ethics is, quite literally, character-forming, but not in any isolationist or purely self-oriented way. Our relations to others call on historical and social traditions that form part of the effective consciousness, the historicity, of the individual, for example specific cultural ideas of justice, right, good, and so on.

Ethical practices are always primarily 'other' rather than 'self' regarding but as Warnke argues (2002: 86) it 'is not possible to give sound [ethical] advice unless one takes the situation to be one that affects one's own life and self-understanding. ... Our understanding of what we ought to do in any situation is not the objective knowledge of an observer but the engaged understanding of someone who must act. ... Questions of what we should do as either individuals or communities are not, Gadamer suggests, questions to be answered by appealing to fixed foundations. They are rather questions that require us to consider who we already are and who we want to become.'

This process of ethical becoming, of forming a sense of appropriate action and of one's place in a community cannot operate unless the individual takes a stance of 'dialogic openness' (Warnke 2002: 93) to the other, whether that be another time, tradition, community or person. Understanding and ethics alike both require that we relate to the other 'in such a way that it has something to say to me' (Gadamer 1998a: 361). This 'something' might be something quite unexpected, something previously beyond our experience that broaches our horizons and questions our previously held understandings. Just as ethics is something different from following rules or orders, understanding is radically different from the simple acceptance of pre-determined ideas. Both require active engagement on the part of the individual and a willingness to challenge the prejudices that compose our current situation. Both require us to be disposed to, or forced to, allow something other to effect/affect us. Of course, from this perspective, Luc Ferry's (1995) description of the modern mind is precisely one of pre-determined closure to the alien otherness of nature, a nature that is always already regarded as a dead-letter.

Hermeneutics and Environmental Ethics

Now, as I have already indicated, there are problems with straightforwardly applying Gadamer's hermeneutics to the natural world. Indeed Gadamer explicitly defines the movement of understanding and ethics, the process of 'becoming' an individual within culture (*Bildung*), as a movement away from nature (Smith 2001). This movement, and our understanding of others and ourselves, takes place in a conversation 'that, however various its languages, always takes place in human, learnable ones. Man [sic] "has" the word ... and that is precisely what distinguishes him from all other natural creatures' (Gadamer 1998b: 4). 'Because we [humanity] are a conversation, we are the one story of mankind [sic]' (Gadamer 1998b: 3). Human culture is precisely a 'sharing with' each other or *Mitteilung* (Gadamer 1998b: 6) occurring through human language. How then can we possibly begin to hear nature speak?

Gadamer's account of the hermeneutic gulf between culture and nature raises at least two serious problems for any environmental ethics. First, Gadamer assumes that any determinate meaning that natural things might have must be found and spoken of (expressed) within the enclosed space that is human conversation and culture. 'All understanding is interpretation, and all interpretation takes place in the medium of a language that allows the object to come into words and yet is at the same time the interpreter's own language. ... The linguisticality of understanding *is the concretion of historically affected consciousness.* ... the essence of tradition is to exist in the medium of language, so that the preferred object of interpretation is a verbal one' (1998a: 389). Such quotations give the impression that, for Gadamer, expression and interpretation are wholly dependent on *people putting things into words*, into a human language, which is itself, an expression of cultural traditions, of 'effective human historicity'. Though, as I hope to show, this impression is not entirely accurate.

A second problem is that despite his emphasis on the need for attitudes like 'openness', 'tact' and 'sensitivity', and the way his account stresses the ethical connotations of genuine communication, Gadamer still underplays the role of 'feeling' in understanding. To say this will seem odd to those who have criticized him for anti-Enlightenment irrationality in over-emphasizing sub-conscious prejudices and under-emphasizing the role of an independent reason in successful communication. (The two key critics are Habermas and Apel, see McCarthy 1984, 162-93.) However, I think it fair to say that Gadamer's accounts of hermeneutics and ethics are strangely lacking in either motivation or emotion. Why exactly are we expected to appreciate and engage with novel cultural experiences, texts or people? Why should we be interested in self-cultivation, in 'becoming' as well as being? Where is the passion in the ethical relation between reader and text, interpreter and interpreted? Exactly what kind of 'sensitivity' to the other are we speaking of?

Establishing an affective relation through sensual experience is as much a part of understanding as anything delimited linguistically, and is not something extra to, but a part of, understanding. Archimedes' exclamation 'Eureka', 'I have found it', was marked not only by his finding an idea but by the *feeling of elation* that had him rise up and *find himself* running naked from bath to street (see also Bondi, this volume). There are so many texts and contexts where we would have to say someone has not understood the situation if they failed to be *moved* to tears, anger, or laughter, to feel shocked or appalled, if they were not moved, physically, gut-wrenchingly, heart-stoppingly, emotionally.

The linguistic and emotional limitations of Gadamer's theory are clearly part of the effective historical consciousness of modernism alluded to by Ferry. This is indicated by both the dominance of visual metaphors in *Truth and Method*, of a hermeneutic situation (a place) bounded by 'horizons', and in terms of an understanding bounded by (human) language.[2] What is missing here is precisely an account of being(s) *touched* by those that cannot speak and the embodied and emotional nature of human experience. What then are we to say of those situations where the other takes hold of/in us, where the wind caresses our cheek, the birdsong lifts our hearts? What about those places where sensations rush in on us, confound us, challenge or reconstitute our embodied sense of self, shift the boundaries between other(s) and ourselves? What happens when that which is other flows into our affections, overpowers us with breathless wonder, holds us spellbound, puts us in our place or *moves* us to another('s) place?

To account for such experiences an environmental hermeneutics must de-centre both anthropocentric aspects of Gadamer's project. It must remove the anthropocentrism present in modernist discussions of the nature of language and the languages of nature. That is, it must recognize that nature too can express itself in ways open to understanding (Smith 2001).[3] But it must also recognize the link between the nature of our feelings and our feelings for nature, the relation between sensuality (being able to feel) and sensibility (having a feeling for) what is appropriate in 'natural' places. This requires that we reconfigure our understanding of what Gadamer refers to as the 'structure of experience', that which is recognized as the background against which and within which all interpretation takes place.

This reconfiguration must recognize the interpretative role of our experiences of nature as well as culture, and of embodied feelings as well as abstract reasons. The structure of experience must refer not only to 'historically effective consciousness' (*wirkungsgeschichtliches Bewusstsein*) but to an 'affective natural history'.

Wittgenstein, Gadamer and 'Natural History'

Interestingly, this de-centring proposal need not entail as radical a shift in Gadamer's problematic as might initially seem to be the case. There are aspects of Gadamer's work that can justify both an increased emphasis on affective aspects of hermeneutics and the inclusion of nature's non-human expression, the languages of nature. One way of illustrating this is through a brief comparison with Wittgenstein's later work, which Gadamer admits, has many similarities to his own (Gadamer 2001: 56).

Like the later Wittgenstein, Gadamer regards understanding as an achievement in contextualizing a word, sentence, poem, piece of music and so on, to see how it relates to and fits within the specific situation in which understanding is taking place. Like Wittgenstein, this understanding is not an abstract objective conceptual process but a calling upon one's previous experiences so as to feel/know how the *sense* of what is expressed is similar to and differs from what has gone before. 'We speak of understanding a sentence in the sense in which it can be replaced by another which says the same, but also in the sense in which it cannot be replaced by any other' (Wittgenstein, 1981: §531). Understanding involves feeling that, and knowing how, something seems to fit into *place*. Once we understand something we are able to relate to the situation in a different way. Understanding is, quite literally, the feeling/knowledge of how things *make sense* to us. Wittgenstein places far more emphasis than Gadamer on the affective nature of human expression and the interpretative importance of linking meaning and understanding to feelings. 'But when one says, "I *hope* he'll come" – doesn't the feeling give the word "hope" its meaning? ... The feeling does perhaps give the word "hope" its special ring, that is, it is expressed in that ring' (Wittgenstein, 1981: §545). For Wittgenstein words can be 'charged with desire' (Wittgenstein, 1981: §546).

Interestingly Wittgenstein also argues that meaningful expression, that is, expression that can be understood, is not limited to human *words*, 'Can I not say: a cry, a laugh, are full of meaning? And that means roughly, much can be gathered from them' (Wittgenstein, 1981: §543). Nevertheless, Wittgenstein famously places an anthropocentric limit of what we can understand when he states, 'If a lion could talk, we could not understand him' (Wittgenstein, 1981: 223). In other words Wittgenstein believes that because we have so little in common with lions then their expressions would fall outside of what Gadamer would refer to as our 'structure of experience' and would therefore be incomprehensible to us. In short, we might say that understanding is being able to fit what is expressed to us by others with (though not necessarily within) our structure of experience. This is what Gadamer refers to as a fusion of horizons. Interpretation just is this process of trying to make things fit sensibly and sensitively into place.

Of course Wittgenstein does not refer to a structure of experience but in several places Gadamer actually appropriates Wittgenstein's phrase 'form of life' as a synonym for the practices within which the structure of experience develops. This seems eminently suitable since the form of life is what is 'given' prior to interpretation, it is the fore-structure of understanding. 'What has to be accepted, the given, is ... so one could say ... *forms of life*' (Wittgenstein 1981: 226).

The real point of contention for an environmental hermeneutics is then whether the non-human does or could constitute an aspect of our structure of experience, because only then will it be able to speak to us in any understandable way. And here, to some extent, I would want to differ from Wittgenstein. I have argued elsewhere that it is only because of a prejudice peculiar to post-Cartesian modernism that we refuse to recognise that many non-human things, including lions, can and do express themselves intelligibly, that they do have 'language' in the broader sense recognised by Walter Benjamin (1998; Smith 2001). Interestingly, Gadamer too has sometimes uncharacteristically admitted that we can 'speak not only of a language of art but also of a language of nature – in short, of any language that things have' (Gadamer, 1998: 475). This admission comes near the end of *Truth and Method* where he is explicating the basic ontological structure of language, that is, the idea of language as the communicable expression of being(s), as an activity of things. Language can be seen, Gadamer says, as 'something that the thing itself does and which [human] thought suffers' (474). Language is then another name for those expressions of things that are able to make themselves understood, to *spring up* into and alongside that which composes our structure of experience. 'Being that can be understood is language' (474).

And, while Wittgenstein might deny that lions – and certainly springs – have anything to say to us that we might understand, he is not so closed minded when it comes to explicating the idea of a form of life. The form of life is not defined or delimited by social practices alone, it isn't just a matter of 'effective human consciousness'. In fact, Arnswald argues, this 'encompassing background is constituted by what Wittgenstein terms *natural history* (*Naturgeschichte*)' (Arnswald 2002: 26).[4] By this I think he means that this ability to understand certain things seems to come naturally to us because of who *and what* we are. 'What we are supplying are really remarks on the natural history of human beings; we are not contributing curiosities however, but observations which no one has doubted, but which have escaped remark only because they are always before our eyes' (Wittgenstein 1981: §415).

The form of life that is the basis for all meaningful interpretations is manifold, and we grasp its many varied forms and manifestations only fleetingly through that experience of understanding itself. It is a mysterious movement that we cannot pinpoint, like that of clear water welling up in clear water. It seemingly comes from nowhere, but its hidden source is all that has already gone to make us up – our nature *and* our culture, and it holds within itself that which we have the possibility to become. And strangely, Wittgenstein's preferred metaphor for the structure of experience which facilitates understanding was that of a river-bed. The 'river-bed of thoughts may shift. But I distinguish between the movement of waters on the river-bed and the shift of the bed itself; though there is not a sharp division of one from the other. ... And the bank of that river consists partly of hard rock, subject to no alteration or only to an imperceptible one, partly of sand, which now in one place now in another gets washed away, or deposited' (Wittgenstein 1988: §97-9). Understanding and interpreting is a precarious occupation, which springs itself upon us, threatening to reconstitute our structure of experience, rewrite our emotional geographies. We may try to orient ourselves by those things that remain relatively fixed but nothing stays still. Expressing one's being (existence) and becoming someone different means opening ourselves up to understanding, to feelings and knowledge. Only then might we try to speak of those things that would otherwise be denied a voice, only then can we begin to write of the Earth as a source of emotion and a subject of concern.

Notes

1   So far as I am aware no one has made this connection in meanings between the River Manifold and the religious and philosophical connotations associated with the term. It is usually thought to derive more prosaically from 'many-folded' in reference to its winding its way though the valley. But the river is not especially folded – and is virtually unique in the scale of its swallow holes and springs. As for the River Dove locals pronounce this as in trove, not as in 'dove', the bird. This too seems to have ancient origins. 'Dove, in so far as it is a genuine old river-name, means "the dark river" and is derived from the Celtic adj. dubo – "black, dark" [...] The Doves do not all have dark water, but they have at least a dark bed or they run through a deep valley' (Ekwell (1928).

2   'Every finite present has its limitations. We define the concept of "situation" by saying that it represents a standpoint that limits the possibility of vision. Hence essential to the concept of situation is the concept of "*horizon*" ... working out the hermeneutical situation means acquiring the right horizon of inquiry for the questions evoked by the encounter with tradition' (Gadamer, 1998: 302). On the prevalence of visual metaphors in modernity see Levin (1993).

3   And here we must recognize that I am not talking about a scientific understanding which is only one way in which nature can be forced to speak to us, to reveal itself, and as Heidegger argues, a way that is instrumental rather than sensual or ethical. Feelings and values are precisely what the scientist claims to leave behind in order to gain objectivity.

4   Wittgenstein was by no means alone in trying to re-appropriate this term. As Hanssen (1998: 3) argues, Walter Benjamin's 'positive validation of natural history was meant to overcome the limitations of historical hermeneutics, whose category of "meaning" (*Sinn*) remained grounded in the understanding of a human subject'.

References

Arnswald, Ulrich (2002) 'On the Certainty of Uncertainty: Language Games and Forms of life in Gadamer and Wittgenstein', in Jeff Malpas, Ulrich Arnswald and Jens Kertscher (eds.) *Gadamer's Century: Essays in Honor of Hans-Georg Gadamer* Cambridge Massachusetts: MIT Press.

Benjamin, Walter (1998) 'Of Language as Such and the Language of Man', in *One-Way Street and Other Writings*. London: Verso.

Ekwall, Eilert (1928) *English River-names* Oxford: Clarendon.

Ferry, Luc (1995) *The New Ecological Order* Chicago: University of Chicago Press.

Gadamer, Hans-Georg (1998a) *Truth and Method* New York: Continuum.

Gadamer, Hans-Georg (1998b) 'Culture and the Word', in *Praise of Theory. Speeches and Essays* New Haven: Yale University Press.

Gadamer, Hans-Georg (2001) *Gadamer in Conversation: Reflections and Commentary* Richard E. Palmer (ed.) New Haven: Yale University Press.

Hanssen, Beatrice (1998) *Walter Benjamin's Other History. Of Stones, Animals, Human Beings and Angels.* Berkeley: University of California Press.

Levin, David (ed.) *Modernity and the Hegemony of Vision* Berkeley: University of California Press.

McCarthy, Thomas (1984) *The Critical Theory of Jürgen Habermas* Cambridge: Polity.

Smith, Alan (1995) *The Manifold Valley and Ilam* unpublished manuscript.

Smith, Mick (2001) 'Lost for Words? Gadamer and Benjamin on the Nature of Language and the "Language" of Nature', *Environmental Values* 10: 59-75.

Warnke, Georgia (2002) 'Hermeneutics, ethics and politics', in Robert J. Dostal (ed) *The Cambridge Companion to Gadamer* Cambridge: Cambridge University Press.

Chapter 17

The Place of Emotions in Research: From Partitioning Emotion and Reason to the Emotional Dynamics of Research Relationships

Liz Bondi

Introduction

As this volume testifies, emotions are currently an important focus of research in several social science disciplines. And yet, as Rebekah Widdowfield (2000) has recently noted, there remains considerable reluctance to discuss the emotional impact of research on researchers themselves, at least in print. There are, of course, exceptions (see for example Hunt 1989; Meth with Malaza 2003; Parr 1998; Laurier and Parr 2000; Wilkins 1993; Young and Lee 1996), but it is probably fair to say that researchers' emotions are spoken about in numerous informal conversations, such as those that follow seminars and conference papers, to a disproportionately greater extent than is acknowledged in published accounts of research. Moreover, in so far as researchers' emotions are explored in print, discussion tends to be limited in terms of both the kinds of research and the range of feelings included.[1] With respect to the former, Rebekah Widdowfield (2000, 201) is typical in linking the relevance of consideration of researchers' emotions to the use of qualitative research methods, which bring 'researchers into direct contact with their research subjects through for example, interviews, ethnographies, and life histories'. With respect to the latter, Elizabeth Young and Raymond Lee (1996, 111) argue that '[t]he emotions expressed in fieldwork accounts tend to be negatively cast, or they express difficulties which are finally managed', while Eric Laurier and Hester Parr (2000, 99) assert that anxiety is 'the classic interviewer's emotion'. Anxiety certainly exudes from several published accounts (not only those that discuss interviews), such as Ruth Wilkins' (1993) discussion of researchers' emotions as sensitising and interpretive resources, Kim England's (1994) reflections on abandoning a research project, and Hester Parr's (1998) exploration of methodological aspects of research about psychiatric service users.

This chapter challenges both limitations and argues for a wider appreciation of researchers' emotions in research practice. I begin by illustrating the ubiquity of what sociologist Arlie Hochschild (1983) has called 'emotion work' in academic research. I then explore the place of emotions in traditional scientific epistemologies, through which I elaborate an understanding of emotional life as itself ubiquitous. Turning to research that generates data by means of interpersonal interactions between researchers and research participants, I argue that emotions are integral to research relationships, and I draw attention to the wide range of emotions experienced by researchers in response to these relationships. Against this background I use an example drawn from my own doctoral research to illuminate practical, methodological and substantive aspects of the place of researchers' emotions in research.

On the Ubiquity of Emotion Work in the Practice of Research

Despite the neglect of researchers' emotions in published discussion, I would suggest that it is actually commonplace, unremarkable and routine to raise questions informally and often implicitly about the 'emotion work' entailed in conducting research. This may be most obvious in relation to 'apprentice researchers', that is candidates for research degrees working with research supervisors.[2] Supervisors often ask students how they feel about immersing themselves in a particular topic, or about key phases of research such as beginning and ending fieldwork. Such questions are rarely formalised, and may be understood by both parties more in terms of pastoral care or support than in relation to academic concerns. Enquiries of this kind are likely to be prompted by more or less overt awareness that it is much easier for students (and other researchers) to work steadily and productively if their projects elicit feelings of excitement, pleasure and personal meaning at least some of the time, than if they generate persistent feelings of boredom, alienation, frustration, inadequacy or anxiety. Related concerns animate conversations among peers within and beyond the academy. For example, friends sometimes ask what drives researchers to spend so much time on a particular topic, and in these informal contexts researchers are likely to convey feelings of attachment, fascination, passion and so on (for a powerful published statement see Metcalfe 1999).

   While such feelings seem necessary to account for the single-minded and sustained focus required for much academic research, novice and experienced researchers also often find themselves faced with questions about their capacity to look at issues and evidence 'objectively' or with 'an open mind'. Even among researchers who criticise claims to objectivity in research, and who understand their work in terms of the situated production of knowledges, the capacity to reflect critically, to think afresh, and to advance arguments capable of convincing sceptical audiences is accorded considerable importance (Haraway 1988). Putting this another way, researchers informed by different epistemological traditions are all expected (by themselves and others) to move between different positions in relation to their work, and these various positions and relationships to research are emotionally inflected. The emotions expected of researchers thus range from the passionate immersion associated with the 'drive' needed to conduct research, to the cool contemplation associated with the capacity to 'stand back' and reflect critically on one's own ideas.

Arlie Hochschild's extensive and highly influential work on emotional labour and regimes of emotion began with a concern about the commercialisation of human feeling (Hochschild 1979, 1983). She laid bare how 'performing' particular emotions has become an increasingly integral component of many jobs, especially in the service sector, and how profoundly such work impacts on the emotional lives of workers. While her account of *The Managed Heart* (Hochschild 1983) emphasised the personally intrusive quality of much emotional labour, she has increasingly relinquished any concept of an authentic, untouched, emotional life, for a more fully socialised and socially constructed view of the ubiquity of 'emotional regimes' in all human life (Hochschild 2002). Although the emotional dimensions of academic labour may be less problematic, commercialised and exploitative than those associated with much service sector work, academic researchers are nevertheless also called upon to manage how they feel and to undertake emotion work, within the context of specific 'emotional regimes'.

In this context, I offer two brief examples of the management of emotions associated with kinds of research generally neglected in the small literature on the emotional impact of research on researchers. Contrary to Rebekah Widdowfield's (2000, 201) assertion that 'face-to-face contact' with '"real" people' generates 'much more intense feelings' than 'numbers', I would argue that quantitative research may be just as emotionally engaging and demanding as qualitative research, albeit in different ways. Drawing on my own experience of both forms of research, while different kinds of feelings are elicited in different contexts, I have certainly experienced frustration and delight of great intensity when struggling with computer software and numerical computation (for an example of the research concerned see Bondi 1991). Indeed, controlling flashes of rage that threaten to overwhelm the capacity to continue with a task such as diagnosing and correcting errors in a computer model, is, surely, a very clear example of the kind of emotion work performed by quantitative researchers. Similarly, archival research is well known to elicit powerful emotions in researchers, whether connected to the sensory experiences of working in archives or with original documents, or through efforts to imagine and understand the lives and legacies of those they research. Here too, researchers are expected to moderate and manipulate their feelings, perhaps by controlling sensory pleasures, cultivating scepticism in place of imaginary (over-) identification, or, conversely, quelling distaste in favour of empathic understanding.

The preceding examples illustrate ways in which researchers are called upon to control the impact, and counteract the intensity, of some common emotional responses associated with research practice. They also help to illustrate the ubiquity of emotion work within research. It does not follow from this that researchers' emotional experiences should necessarily become the subject of academic scholarship, but it does suggest that reflections on such experiences could usefully be reframed in more inclusive ways. I develop this argument in greater depth by considering the place of emotions in the archetypally dispassionate realm of research in the natural and physical sciences.

Emotions and Science

In the popular imagination, scientific research is objective rather than subjective in the sense of being grounded in evidence that lies outside the internal world of the researcher and that is available to scrutiny by others (see, for example, Chalmers, 1999). In other words, researchers are assumed to be, and to remain, detached from their evidence. Popularly, science is also understood to be grounded in rationality and logical argument rather than subject to the whims, preferences or pre-determined judgements of researchers. Analysis proceeds logically with conclusions traceable back to the evidence through explicit rules of interpretation. Values are held in check and, classically, science is presented as value-neutral.

Emotions are widely understood to 'interfere' with the capacity for logical reasoning, to 'cloud' the judgements required to apply rules of interpretation systematically, and to produce inappropriate 'biases'. This suggests that emotional life has no place at all in scientific research. But the common assumption that emotions are antithetical to, and have no place within, science, is easily contradicted: scientific research is also popularly imagined to depend upon moments of inspiration and creative thinking, which are understood to entail intense emotions. Perhaps most famously, Archimedes' exclamation of 'Eureka/I have found it' when he realised how the displacement of water would enable him to work out the volume of an irregular solid, illustrates the enormous excitement associated with, and expected of those involved in, major scientific breakthroughs. The story goes that Archimedes' breakthrough came about not through the application of rational, logical thinking, but through pure inspiration. The inspirational moment came as Archimedes stepped into the bath, and, realising in a flash the solution to the problem on which he had been working, he is reputed to have felt so excited that he ran through the streets of Athens naked, exclaiming 'Eureka!' (also see Smith in this volume). And so the reputation for eccentricity among scientists is traceable to Archimedes. I would suggest that eccentricity attaches to scientists at least partly because scientific research depends upon two sides of a supposedly mutually exclusive dichotomy between the dispassionate detachment associated with rationality and objectivity, and the deeply creative immersion (literal or figurative) associated with moments of great insight and excitement (also see Ghiselin, 1952; Metcalfe 1999).

If they acknowledge emotional dimensions of scientific research at all, scientists working within positivist or critical rationalist traditions typically argue for a strict separation between the non-rational (including the inspirational and emotional) and the rational. For example, Karl Popper (1972) has argued that inspiration and creativity are essential parts of theoretical advancement in science, especially in relation to what he calls 'conjecture'. However, he insisted on a sharp separation between such aspects and the application of the scientific method through rigorous attempts at refutation. Thus, for Popper, non-rational aspects of life, such as emotions, do have a place in the working lives of researchers, but a place firmly separated from the application of logic and rational argument, for example, in processes of hypothesis testing, experimental design, theoretical development and so on. More generally, scientific epistemologies insist on the demarcation of scientific from non-scientific knowledge, logic from intuition, and rationality from emotional life. Thus, within such epistemologies, the place of researchers' emotions in research is a firmly demarcated one, unambiguously separated from logic, rationality and objectivity.

This strict separation is challenged by post-structuralist critiques of science (see, for example, Hekman 1990). Post-structuralism shows how ideas about objectivity and rationality depend upon an implicit dissociation from, and claim of superiority to, what they are not. On this account the underlying dissociation does not rest on neutral, logical foundations, but is a strategy to claim authority and power. For example, from a post-structuralist perspective, objectivity presupposes what it excludes, namely subjectivity, because objectivity depends upon human subjects exercising their subjective capacities. A clear distinction between objectivity and subjectivity does not, therefore, stand up to close inspection, and proclamations of scientific objectivity are primarily about asserting the authority of certain kinds of knowledge. Similarly, rationality takes its form only by the negation of what is not rational; it depends on the idea that two domains can be framed as mutually exclusive. Moreover the non-rational, or irrational, is often associated with emotional life, with the feminine, and with the body. But there is no logical basis for the assumption that rationality can be separated from its opposite, or that mind can be exercised without body. It may be convenient to think like that, but it is not, in fact, 'rational' (Rose 1994; Williams and Bendelow 1998).

By deconstructing mutually exclusive or binary oppositions between rationality and non-rationality, and between objectivity and subjectivity, this post-structuralist critique questions the idea that scientific knowledge can be distinguished from non-scientific knowledge. One consequence of this is to relativise all claims to knowledge: if there is no absolute or universal basis to scientific knowledge then all that can be said is that some forms of knowledge are more (or less) convincing, successful, useful or influential than others. Many issues flow from this, including what, within a post-structuralist perspective, constitutes rigorous, well-formulated research. It is not my purpose to address such issues directly. Instead I want to draw out just one point. If rationality and objectivity are infused with their supposed opposites, then we might usefully think of states like 'detachment' as themselves emotional. Likewise a state of dispassionate, cool contemplation can be thought of as requiring a certain kind of emotional commitment, or, as Philip Rieff (1960) puts it, an 'irrational passion for dispassionate rationality'. Put another way, the practice of logical thought, the exercise of self-restraint, and forms of intellectual absorption that foster a sense of separateness and detachment, as well as a passion for sharp demarcations, may be deeply pleasurable, or at least preferable to alternative ways of being. As feminist theorists have argued, such preferences are bound up with ideas and experiences of gender (Lloyd 1984). Deconstructing the binaries associated with scientific epistemologies therefore calls into question the association between femininity and emotion, as well as drawing attention to disavowals of emotion associated with the pursuit of 'reason'.

On this account, emotions are not just 'hot' feeling states such as the thrilled excitement mythologised in accounts of Archimedes' bath-time discovery but can usefully be considered to include a much wider panoply of feelings, including those of coolness and restraint. If emotions are understood in this way, it follows that thinking is never emotion-free: rather, our feeling states and our thinking are closely intertwined. Thus, the partitioning of emotion and reason associated with scientific epistemologies can be recast as a form of emotion work that demands the radical separation of (what may be pleasurable feelings of) detachment, objectivity, logic and so on, from other kinds of emotional experiences.

It is not the purpose of this chapter to explore the implications of this analysis for scientific research. Instead I want to use it to inform a discussion of researchers' emotions in forms of qualitative research that depend upon researchers' capacities to forge effective and appropriate interpersonal relationships with others, variously called research subjects, participants, informants or respondents.

Emotions and the Co-construction of Data in Interpersonal Relationships

Gathering or generating data always draws researchers into relationships. These might be relationships with texts, with numerical datasets, or with the interfaces through which such data are extracted or analysed, all of which, as I have already argued, evoke emotional responses in researchers. Nevertheless, research methods that draw directly upon interpersonal interactions, such as interviews and participant observation, require researchers to use themselves in distinctive ways since the people with whom they interact are also sentient, feeling human beings. Data generated by such methods are not so much collected as produced or constructed or co-constructed (see for example Limb and Dwyer 2001; Mason 1996; May 2002). Evidence, in the form of tape-recorded conversations or fieldnotes or other formats, is not simply given to researchers by the people with whom they interact, but comes into existence through the interaction itself. Both parties are actively involved in the creation of data in the course of their various interpersonal encounters, and these encounters are rich with emotions and emotional dynamics.

The co-construction of data in interpersonal relationships requires both researchers and those with whom they interact to deploy a wide range of skills to which emotional life is integral. Although emotions are often popularly understood as 'private' or 'internal', they are also understood to entail movement across the boundaries that constitute 'private' or 'internal' realms, for example, when attributed to 'external' stimuli, and in their scope to be communicated to others. Indeed people are highly sensitive and responsive to one another's feelings, something that is possible only because aspects of emotional life traverse, and can be sensed across, boundaries between the supposedly 'internal' lives of different people (Chodorow, 1999). Much of the time these emotional dimensions of research relationships remain taken-for-granted, unnoticed and out of conscious awareness. This poses numerous challenges to scholars in their attempts to make emotions the focus of research. According to Ian Craib (1995) much of the sociological literature on emotions fails to engage adequately with these challenges, leading to overly cognitive accounts of emotional life, and neglect of the importance of unexpressed and inexpressible feelings as well as conflicts between thoughts and feelings. Nevertheless, at least some aspects of the emotional dynamics of research relationships can usefully become the subject of reflection and analysis.

One characteristic feature of emotional life is the mutability, fluidity and multiplicity of feelings. For example, we may (as researchers) approach a research interview with trepidation, anxiety and/or determination; become so immersed in the encounter itself that we lose all track of what we are feeling (see Bondi 2003); leave feeling pleased, relieved, and/or relaxed; and subsequently find ourselves worrying about how the interviewee might be feeling, feeling guilty, and/or longing for an opportunity to meet our interviewee again. These feelings may barely register, we may seek to ignore some or all of them, and/or we may consciously focus on one or more of them. Perhaps the rich mutability and complexity of emotional life is at its most obvious if we attempt to imagine its absence: if researchers remain unremittingly and singularly excited, anxious, numb or detached, we would be disconcerting and off-putting people with whom to interact in all but the most fleeting encounters. Instead, the interpersonal interactions that constitute many forms of qualitative research evidence are likely to evoke complex, rich and fluid mixtures of such feelings as wariness, excitement, anxiety, pleasure, boredom and so on. Most of the time researchers' changing emotional experiences remain taken-for-granted and largely unnoticed, attracting attention and becoming 'noteworthy' only if the researchers' ordinary 'flow' of feelings is interrupted, for example, by persistent and heightened anxiety (see Wilkins 1993), intense fear (Meth with Malaza 2003), or emotionally generated nausea (Johnson 1975; cited by Young and Lee 1996, 99).[3]

Drawing on Arlie Hochschild's (1983) theorisation of emotional labour, Elizabeth Young and Raymond Lee (1996) argue that different traditions of qualitative fieldwork emphasise different 'feeling rules', and that fieldworkers' feelings are most likely to attract attention when researchers are struggling to find or apply appropriate feeling rules (see Hepworth, this volume, for discussion of the concept in a different context). On this account, the feelings that researchers discuss provide clues as to the feeling rules in operation. For example, Rebekah Widdowfield (2000, 205), describes her feelings of horror and upset when first visiting a very disadvantaged and unpopular public housing estate, and she conveys a powerful sense of struggling to work out 'what to do' with such feelings, thereby suggesting a search for guidance to enable her to judge the appropriateness of how she feels. This analysis also sheds light on links between ethics and emotions emphasised in a number of recent discussions (Laurier and Parr 2000; Meth with Malaza 2003; Valentine 2003): if researchers' emotions attract attention in the context of normative judgements about how one should feel, it is not surprising that they prompt consideration of ethical questions.

This theorisation of researchers' emotional experiences is useful in highlighting unexamined normative assumptions about how researchers should feel. Although not explicitly framed in terms of an analysis of 'feeling rules', Ruth Wilkins (1993) illustrates a version of this when she chose to contest advice 'not to take it personally', that is when she refused to conform to a feeling rule that required her to manage her fear of rejection and acute anxiety by quelling, minimising and controlling such emotions. Identifying feeling rules also helps to call into question the attribution of 'negative' or 'positive' values to particular emotional states since such evaluations simply express normative expectations associated with feeling rules and emotional regimes. For example, Rebekah Widdowfield (2000, 205) describes feelings of horror as 'negative' without any qualification. But the 'negative' character of such feelings is by no means universal as the pleasure audiences take in watching 'horror movies' indicates. In other words, Widdowfield's labelling of horror as negative reflects her sense of transgressing implicit feeling rules associated with her fieldwork. However, while the concept of 'feeling rules' may help to highlight the circulation of normative expectations about researchers' emotions, it risks missing other important issues, as I seek to illustrate in the next section of this chapter.

Learning from Mixed Feelings

This section draws on my own experiences as a doctoral researcher some 20 years ago. It begins with what was a noteworthy, but then un-noted and unacknowledged experience of discomfort. My exploration here seeks to draw out something of the layering of emotions, and to consider some of the implications that ensue.

During the course of my doctoral research I conducted interviews with people who were involved in local campaigns against plans to close or merge primary schools in a British city. I was interested in hearing about people's experiences, and there was much about the interviews that I enjoyed. However, one feature of the fieldwork troubled me a good deal. I would always explain, at least once, and often more than once, that neither my research nor I would influence the outcome of the process of rationalising primary school provision in which the local education authority was engaged. Schools would close, and participating in the research would in no way affect which schools or how many. But despite saying this, I still kept getting the sense that quite a few people went on hoping that I was willing to be an advocate for them, and that I had the power to make a difference.

As the fieldwork continued I felt more and more uncomfortable about this feature of my experience. I felt guilty because, despite my efforts to be honest, people were spending time with me and were telling me all about their experiences, perhaps at least in part, on the basis of a misapprehension about my capacity to offer anything in return. This played into an underlying feeling, which I came to think of as characteristic of my experience as a doctoral candidate, and which I gradually discovered was and is by no means unique. I felt fraudulent and always at risk of being 'found out' or 'found wanting', which went along with a belief or a fantasy that I had progressed through the academic system to a PhD studentship because of a mistake rather than because of my abilities. I did not think I was 'good enough'. However, I did not admit these feelings to anyone at the time. They are relevant to my account now because they illustrate how the feelings of guilt prompted by the research interviews I conducted became interlaced with, and intensified by, my much more general sense of inadequacy. Both guilt and inadequacy are often associated with shame, and, although I would probably not have recognised shame as a feature of what I felt at the time, it now seems to be a very apt description of the feelings I seek to convey.[4]

Feelings of shame are strongly associated with the urge to hide, which is suggested by my not disclosing how I felt. But my sense of shame and impulse to hide should not be over-generalised or over-stated. I did not seek to hide all aspects of myself, and I enjoyed many aspects of my research, including the self-same interviews that prompted my feelings of guilt. However, I did feel compelled to hide my sense of inadequacy and my sense of guilt, hence my conviction that I did not (consciously or intentionally) disclose these feelings to anyone – not to my supervisors, peers, friends or family.

Feelings of guilt, shame and inadequacy are often debilitating, and can become 'paralysing' in the sense of seriously impeding or even preventing the continuation of research (England 1994; Widdowfield 2000). In my own case, like many of us much of the time, I was able to exercise sufficient control over these feelings to complete my fieldwork, write and defend my thesis, and move on to new research.[5] Perhaps, therefore, these feelings can safely and appropriately be considered irrelevant to the research in question and consigned to the limited period of doctoral training. However, I think that this would be to neglect three potentially important issues raised by this account, which relate to practical, methodological and substantive aspects of research respectively.

The first, practical, concern is about how, and to what extent, research communities might attend to the emotional impacts of research on researchers themselves. As a doctoral student, I did not expect or seek support in relation to such impacts. Indeed I had no conceptual framework through which to name my emotional experiences as part of, or relevant to, my research. Moreover, as I have said, I was able to control potentially debilitating feelings so that they did not in any obvious sense impair my progress. But is it enough to assume that researchers, whether students or not, can manage in this way? Researchers working on topics that they themselves, or members of their research teams, construe as likely to be emotionally demanding sometimes build in arrangements for support (see for example Bingley 2002; Burman and Chantler 2004; Young and Lee 1996). On other occasions it may be appropriate to decide that one is not the right person to conduct the research in question (England 1994). However, my question is about those far more numerous examples of research that may have all kinds of unexpected and unanticipated emotional impacts on researchers (including work that does not include direct interpersonal interaction) such as my own doctoral research.

One possible response would be to ensure that all researchers have clearly established access to people willing, able and trained to reflect with them on such impacts within appropriate boundaries of confidentiality. In some cases doctoral supervisors or colleagues might be able to offer such opportunities, but in other cases students may not feel safe and secure enough with supervisors to talk openly about their emotional experiences, and the same may apply to more established researchers in relation to their colleagues. In such circumstances a greater degree of 'distance' may be needed in order to ensure sufficient confidence as well as confidentiality to enable constructive reflection to take place. Conversely supervisors and colleagues may not feel willing, able or adequately trained to listen to how other researchers feel, perhaps especially if they have never had, or availed themselves of, such opportunities themselves. They may also experience conflict between responding supportively to emotional disclosures and other aspects of their responsibilities, such as ensuring the timely completion of doctoral students' theses. To make access to confidential opportunities for such reflective work real and effective is therefore likely to require thoughtful planning and investment in training. In practice, many researchers, maybe most researchers most of the time, would not wish to take advantage of access to such opportunities. But their availability could signal an awareness of emotional impacts that might itself facilitate informal supportive reflection. In other words, a modest institutional investment in this area could yield wide-ranging benefits.

It might be tempting, and in some cases appropriate, to link together opportunities to reflect on emotional issues and ethical questions. But the example I have used suggests that caution might be needed. I felt guilty about what I was doing as an interviewer and I felt that I must be 'doing something wrong'. To have approached the issues at stake in terms of research ethics would not, I believe, have been very helpful to me. I was already working within appropriate ethical protocols, and the 'problem' was not about what I was *doing* but with how I was processing what I was *feeling*. Had I been able to make use of formal opportunities to reflect on what I felt, I think that I would have benefited far more from a strongly non-normative approach than one too closely tied to explicit ethical considerations. I needed, at least initially, to 'accept' my feelings of guilt, rather than to try to prevent them from happening. Unacknowledged shame and my more general sense of inadequacy would have needed to come into my awareness for such acceptance to be possible. Only with such acceptance could I possibly have begun to reflect on what my feelings of guilt might mean in relation to the substance of my research.

My second concern follows from this, and relates to the idea that researchers' emotions can be analysed in terms of 'feeling rules' (Young and Lee 1996). My concern with this approach is twofold. On the one hand, the focus on what fieldworkers think that they ought to feel risks inviting researchers into unduly swift and categorical statements about what they feel and why they feel it. As I have already elaborated, in my own case I was struggling to make sense of what I felt. While this suggests the relevance of 'feeling rules', in the sense that I was trying to work out what I should feel, I was also at risk of being overwhelmed by normative considerations because of my sense of guilt, that is my feeling that I must be doing something wrong. One symptom of the overwhelming quality of my emotional experience was that I could not think about it effectively. Instead what I needed was to do considerably more 'emotion work' in the form of reflecting on and processing feelings with someone else's assistance before my feelings could have become useful guides of any kind. A tendency towards 'paralysis' seems to be a more general risk associated with attending to researchers' emotions (England 1994; Widdowfield 2000), and I think that this is another way of describing this experience of being overwhelmed by particular feelings. As I have already noted, several existing contributions link discussion of emotions with questions of ethics, which risks contributing to the swift application of normative judgements about researchers' emotional experiences. My argument is that it is important methodologically to be able to *suspend* normative judgements about what one feels in order to reflect on emotions in their full richness and complexity, and that this is often difficult to do, perhaps especially, although by no means only, when feelings such as shame and guilt hover in the background.

On the other hand, the notion of 'feeling rules' actively draws attention to normative features of accounts at the expense of the non-normative. In so doing it risks failing to draw upon less noteworthy and therefore relatively unproblematised feelings, including, for example, my 'enjoyment' of the self-same interviews that prompted my feelings of guilt. In this context it is important to note that I enjoyed most of my face-to-face contact with interviewees, and I tended to feel guilty only afterwards. Reflecting on my enjoyment, I would suggest that this related to the pleasure many of those I interviewed took in the apparently simple act of telling their stories: I was drawn into and touched by the evident pleasure of others. This is one example of how research subjects' feelings impact on researchers through the dynamic interplay between them (Bondi 2003).

Telling our stories – narrating events in our lives – is an ordinary and necessary practice in many cultures. Most of us feel the need to do this in relation to many different kinds of events, especially ones in which we are emotionally invested, whether as a result of trauma (Brison 1997) or for more ordinary processes of confirming, sustaining and creating our identities (Giddens 1991). And so it is always worth remembering that if people freely consent to participate in research interviews, they probably really do want to make use of the opportunity to talk that it affords them! This is not meant as a license to disregard ethical protocols, good interview practice, or the limitations of informed consent (Vivat 2002); far from it. Rather, the point is that, having ensured that potential interviewees are appropriately informed, willing to participate, and know that they can opt out, researchers like myself should, surely, respect their capacity to make decisions,[6] which is something I was at risk of forgetting in my preoccupation with my own sense of guilt. Put another way, I did not know better than those I interviewed about the pros and cons of participating in the research, but I was at risk of allowing my feelings of guilt to convince me that 'really' the cons must outweigh the pros regardless of what participants said to me. Reflecting on this experience now, and especially on the enjoyment that remains so easily eclipsed by my more troubling feelings of guilt, I am inclined to think that the opportunity to talk to an attentive listener about their involvement in local protest groups was welcomed and enjoyed by many of those I interviewed because they were, in most cases, talking about their first-time experiences of political protest. Not surprisingly they appeared to take considerable pleasure in describing their political activities to someone who was genuinely interested and in 'listening mode'. Under these circumstances, it is hardly surprising that I enjoyed witnessing their self-narration. However, I was at risk of discounting the pleasure that was being communicated, and of managing the research relationships more in response to my feelings of guilt and shame than what participants were actually communicating to me. In other words, by failing to attend effectively to my own feelings, I was, ironically, undermining my capacity to attend effectively to the emotional experiences of my research participants. Thus, attention to multiple aspects of researchers' emotions is important not only for our capacity to (continue to) do research, but also for how we engage with research participants, and for how we think and feel about what we do.

Thirdly, and again with the benefit of hindsight, I think that my experience of guilt might have been a potentially valuable resource within my research in relation to the substance of my analysis (Bondi 2003; Laurier and Parr 2000; Wilkins 1993; Young and Lee 1996). In my thesis, and in subsequent publications, I wrote about dynamics through which some voices were incorporated within the policy process and others were excluded from it, and I pointed to the social class basis of much of this differentiation (Bondi 1987a, 1987b, 1988). Looking back, I do not disagree with that analysis, but I do think that it would have been considerably enriched had I been able to think about the meaning of my own feelings of guilt. These feelings, I now think, provided pointers to the deployment of moral discourses within the interviews I conducted. To elaborate, those involved in campaigns to oppose school closure, together with the politicians and local officials responsible for the reorganisation plans against which they were protesting, variously claimed that they were 'right' and others were 'wrong'. In other words, they all used moral arguments to support their positions, and these moral arguments distributed right and wrong across different people and different schools, using different criteria. Having explained to interviewees that I was 'neutral' in the sense of having no influence on the process, and that I was talking to different people differently positioned in relation to reorganisation plans, I made myself available as an attentive, affirmative listener. While I posed questions that asked people to reflect in ways that went beyond their rehearsed arguments, I did not actively disagree with what they said. Indeed I really was neutral in the sense that I did not have a firm view about the rights and wrongs of the reorganisation plan as a whole or about its impact on particular schools, and it was only long after conducting the interviews that I began to formulate my own ideas about how the process could have been managed differently (Adler and Bondi 1988). Nevertheless, I was actively recruited into the distribution of right and wrong in the sense that my interviewees (obviously and understandably) wanted to convince me that they were 'right'. Because I did not directly contest their accounts, it could be argued that I allowed people to believe that they had convinced me and 'won' me over, and my guiltiness could be understood as an effect of misleading people in this way. But I do not think that I misled anyone about where I stood (on my neutral ground) and I do not believe that anyone was in fact misled. Rather, I think that my guiltiness is better understood as symptomatic of the largely unstated moral dimensions of the arguments to which I was listening and into which people sought to recruit me.

This example of the potential use of researchers' emotions as interpretive resources adds to a small number of examples discussed by other researchers (e.g. Wilkins 1993; Laurier and Parr 2000; Davidson 2001). It also suggests that researchers probably use their emotional responses far more extensively than they explicitly acknowledge: I could have written about the moral dimensions of the process of reorganising primary school provision without explicitly referring to my experience of guilt. This is not surprising (or necessarily problematic) given the close interweaving between feelings and thinking to which I referred earlier. Indeed it serves as a reminder that researchers can make use of their emotions as resources without necessarily writing first person accounts of how they feel. However, the general silence about researchers' emotions impoverishes discussion about the nature of this deployment.

Conclusion

In this chapter I have argued that emotions are an inevitable and necessary aspect of doing research. Even if researchers work within the framework of positivism or critical rationalism, in which the application of the scientific method is conceptualised as emotion-free, emotional aspects of existence are valued as rich sources of the creativity and inspiration that give rise to ideas that subsequently become subject to the rigours of the scientific method. Moreover, whatever their epistemological framework, researchers are called upon to perform emotion work in relation to their research. In research that involves interactions with other people with whom data are co-constructed, researchers enter into interpersonal relationships that generate rich emotional dynamics. In this context, I have argued that researchers experience and negotiate a more diverse and fluid array of emotions than existing accounts suggest. Against this background I have reflected on some of the feelings associated with interviews conducted during my doctoral research some two decades ago, drawing attention to a series of issues prompted by my memories.

Qualitative researchers are often exhorted to be 'reflexive', meaning that we should reflect on our own position within research encounters (England 1994; McDowell 1992; Rose 1997). Like Rebekah Widdowfield (2000), I would argue that emotional aspects of reflexivity remain neglected in the existing literature, although I have also suggested that informally researchers' emotions do attract attention. Bringing researchers' emotions into the domain of publications could be regarded as no more than an uncritical participation in much wider processes of the public performance of emotions or 'emotionalisation' of culture. However, drawing on some of my own experiences as a doctoral researcher I have argued that there are practical, methodological and substantive reasons why researchers *may* benefit from reflecting on their emotional responses to interpersonal interactions entailed in fieldwork. Practically, I have argued that it would be desirable for all researchers to have access to opportunities to discuss emotional experiences of research in confidence and non-prescriptively. Methodologically, I have argued that reflecting on the rich and diverse qualities of researchers' emotional responses to fieldwork experiences may be important to our continuing capacity to conduct fieldwork, to interact sensitively with research participants, and to develop rich understandings of what it is we do. Substantively, I have illustrated the potential relevance of researchers' emotions as analytic resources. Running across all three of these reasons, I have highlighted the importance of less immediately obvious or apparently unremarkable features of researchers' emotional experiences, as well as more troubling and 'difficult' feelings. In summary, the capacity to reflect on emotional experiences inclusively and non-prescriptively serves to enrich research practices in a variety of ways.

Acknowledgements

My thanks to Karen Nairn and Joyce Davidson for their comments on an earlier draft of this chapter. Thanks also to seminar audiences in Norway, New Zealand and the UK for their responses to oral versions.

Notes

1 I do not differentiate precisely between 'emotions' and 'feelings'. Neither do I explicitly position my analysis within a particular theoretical approach to emotions partly because my intention is to offer an argument that connects as closely as possible to 'ordinary', theoretically hybrid, formulations of experience within the kind of academic environments with which I am familiar. While my account is informed implicitly by both sociological and psychoanalytic theories, it is influenced especially by the refraction of the latter through the practice of counselling (Bondi 1999, 2003).
2 The first draft of this chapter was presented as part of a doctoral training programme, and this setting prompted me to focus especially, but not exclusively, on experiences associated with doctoral research. I hoped that my account would be experienced as validating by at least some of the participants who, I guessed, might recognise points of convergence or similarity with their own experience. My hopes were (more than) realised, and were repeated on subsequent occasions when I used the same examples with a range of audiences, generating what were for me several stimulating and pleasurable conversations with undergraduate, postgraduate and experienced researchers. A textual version is unlikely to have similar effects because the character of emotional communication associated with reading (although by no means absent) is so different from that associated with face-to-face interaction.
3 Researchers' feelings that are not 'noteworthy' (in the sense of being recorded in fieldnotes) may, nevertheless, be represented within research, although they are not usually labelled or analysed as emotions. For example, passion, anger and anxiety, as well as detachment, exude from the pages of at least some academic writing.
4 This account illustrates two features of emotional experience, namely (a) that much of it lies beyond, or at the edges of, conscious awareness, and (b) that any discussion of emotion is mediated by particular interpretive frames. Whether understood psychoanalytically as an unconscious or preconscious emotion, or as a *post hoc* rationalisation of how I felt, shame, together with an unwillingness to acknowledge it, fits very well with how I remember my emotional experience at the time to which my account refers. My choice of words, together with the way in which I structure my account, is undoubtedly influenced by my experience as a client of psychotherapy (after the end of my doctoral studies) and my subsequent training as a counsellor (Bondi 1999).
5 I did not, however, build directly on my doctoral research, and, for many years, I was aware of avoiding research interviews which I thought might reproduce the experience of guilt I have described.

6 Of course, the capacity (and legal entitlement) to make decisions varies and special considerations may apply to research subjects whose decision-making capacities are limited, impaired or in doubt.

References

Adler, Michael and Bondi, Liz (1988) Delegation and community participation: an alternative approach to the problems created by falling primary school roles, in Liz Bondi and M.H. Matthews (eds) *Education and Society* London: Routledge, 52-82.

Bingley, Amanda (2002) Research ethics in practice, in Liz Bondi, Hannah Avis, Ruth Bankey, Joyce Davidson, Rosaleen Duffy, Victoria Ingrid Einagel, Anja-Maaike Green, Lynda Johnston, Susan Lilley, Carina Listerborn, Shonagh McEwan, Mona Marshy, Niamh O'Connor, Gillian Rose, Bella Vivat and Nichola Wood *Subjectivities, Knowledges and Feminist Geographies: The Subjects and Ethics of Social Research* Boulder, Colorado: Rowman and Littlefield, 208-222.

Bondi, Liz (1987a) *The geography and politics of contraction in local education provision: a case study of Manchester primary schools* Unpublished PhD, University of Manchester.

Bondi, Liz (1987b) School closures and local politics: the negotiation of primary school rationalisation in Manchester, *Political Geography Quarterly*, 6, 203-224.

Bondi, Liz (1988) Political participation and school closures: an investigation of bias in local authority decision-making, *Policy and Politics*, 16, 41-54.

Bondi, Liz (1991) Attainment in primary schools: an analysis of variations between schools *British Educational Research Journal*, 17, 203-217.

Bondi, Liz (1999) Stages on journeys: some remarks about human geography and psychotherapeutic practice *The Professional Geographer* 51, 11-24.

Bondi, Liz (2003) Empathy and identification: conceptual resources for feminist fieldwork *ACME: an International Journal of Critical Geography* 2, 64-76.

Brison, Susan J. (1997) Outliving oneself: trauma, memory and personal identity, in Diana Tietjens Meyers (ed.) (1997) *Feminists Rethink the Self* Boulder, Colorado: Westview, 12-39.

Burman, Erica and Chantler, Khatidja (2004) There's no-place like home: emotional geographies of researching 'race' and refuge provision in Britain *Gender, Place and Culture* (forthcoming).

Chalmers, A. F. (1999) *What Is This Thing Called Science?* Buckingham: Open University Press, (3rd edition).

Chodorow, Nancy (1999) *The Power of Feelings* New Haven and London: Yale University Press.

Craib, Ian (1995) Some comments on the sociology of the emotions *Sociology* 29, 151-158.

Davidson, Joyce (2001) 'Joking apart ...': a 'processual' approach to researching self-help groups *Social and Cultural Geography* 2, 163-183.

England, Kim (1994) Getting personal: reflexivity, positionality and feminist research *The Professional Geographer* 46, 80-89.

Ghiselin, Brewster (1952) *The Creative Process: a Symposium* New York: Mentor.

Giddens, Anthony (1991) *Modernity and Self-Identity* Cambridge: Polity Press.

Haraway, Donna (1988) Situated knowledges: the science question in feminism and the privilege of partial perspective. *Feminist Studies* 14, 575-599.

Hekman, Susan (1990) *Gender and Knowledge* Cambridge: Polity Press.

Hochschild, Arlie (1979) Emotion work, feeling rules and social structure *American Journal of Sociology* 85, 551-575.

Hochschild, Arlie (1983) *The Managed Heart* Berkeley California: University of California Press.

Hochschild, Arlie (2002) Emotion management in an age of global terrorism *Soundings* 20, 117-126.

Hunt, Jennifer C. (1989) *Psychoanalytic Aspects of Fieldwork* London: Sage.

Johnson, John M. (1975) *Doing Field Research* New York: The Free Press.

Laurier, Eric and Parr, Hester (2000) Emotions and interviewing in health and disability research *Ethics, Place and Environment* 3, 98-102.

Limb, Melanie and Dwyer, Claire (eds) (2001) *Qualitative Methodologies for Geographers* London: Arnold.

Lloyd, Genevieve (1984) *The Man of Reason* London: Methuen.

Mason, Jennifer (1996) *Qualitative Researching* London: Sage.

May, Tim (ed.) (2002) *Qualitative Research in Action* London: Sage.

McDowell, Linda (1992) Doing gender: feminism, feminists and research methods in human geography *Transactions, Institute of British Geographers*, 17, 399-416.

Metcalfe, Andrew W. (1999) Inspiration *Canadian Review of Sociology and Anthropology* 36, 217-240.

Meth, Paula, with Malaza, Knethiwe (2003) Violent research: the ethics and emotions of doing research with women in South Africa *Ethics, Place and Environment* 6, 143-159.

Parr, Hester (1998) The politics of methodology in 'post-medical geography': mental health research and the interview *Health and Place* 4, 341-353.

Popper, Karl R. (1972) *Conjectures and Refutations* London: Routledge and Kegan Paul.

Rieff, Philip (1960) *Freud: The Mind of a Moralist* London: Victor Gollancz.

Rose, Gillian (1997) Situating knowledges: positionality, reflexivities and other tactics *Progress in Human Geography*, 21, 305-320.

Rose, Hilary (1994) *Love, Power and Knowledge* Cambridge: Polity Press.

Valentine, Gill (2003) Geography and ethics: in pursuit of social justice – ethics and emotions in geographies of health and disability research *Progress in Human Geography* 27, 375-380.

Vivat, Bella (2002) Situated ethics and feminist ethnography in a west of Scotland hospice, in Liz Bondi, Hannah Avis, Ruth Bankey, Joyce Davidson, Rosaleen Duffy, Victoria Ingrid Einagel, Anja-Maaike Green, Lynda Johnston, Susan Lilley, Carina Listerborn, Shonagh McEwan, Mona Marshy, Niamh O'Connor, Gillian Rose, Bella Vivat and Nichola Wood *Subjectivities, Knowledges and Feminist Geographies: The Subjects and Ethics of Social Research* Boulder, Colorado: Rowman and Littlefield, 236-252.

Widdowfield, Rebekah (2000) The place of emotions in academic research *Area* 32, 199-208.

Wilkins, Ruth (1993) Taking it personally: a note on emotions and autobiography *Sociology* 27, 93-100.

Williams, Simon J. and Bendelow, Gillian (1998) *Emotions in Social Life* London and New York: Routledge.

Young, Elizabeth and Lee, Raymond (1996) Fieldworker feelings as data; 'emotion work' and 'feeling rules' in first person accounts of sociological fieldwork, in Veronica James and Jonathan Gabe (eds) *Emotions and the Sociology of Health* Oxford: Blackwell, 97-113.

Index